Practical Bioinformatics for Beginners

From Raw Sequence Analysis to Machine Learning Applications

Practical Bioinformatics for Beginners

From Raw Sequence Analysis to Machine Learning Applications

Edited by

Lloyd Low
University of Adelaide, Australia

Martti Tammi
Tanalink Sdn Bhd, Malaysia

World Scientific

NEW JERSEY • LONDON • SINGAPORE • BEIJING • SHANGHAI • HONG KONG • TAIPEI • CHENNAI • TOKYO

Published by

World Scientific Publishing Co. Pte. Ltd.
5 Toh Tuck Link, Singapore 596224
USA office: 27 Warren Street, Suite 401-402, Hackensack, NJ 07601
UK office: 57 Shelton Street, Covent Garden, London WC2H 9HE

Library of Congress Cataloging-in-Publication Data
Names: Low, Lloyd, editor. | Tammi, Martti (Martti T.), editor.
Title: Practical bioinformatics for beginners : from raw sequence analysis to machine learning applications / edited by Lloyd Low, University of Adelaide, Australia, Martti Tammi, Tanalink Sdn Bhd, Malaysia.
Description: First edition. | New Jersey : World Scientific Publishing Co. Pte. Ltd, [2023] | Includes bibliographical references and index.
Identifiers: LCCN 2022022559 | ISBN 9789811258985 (hardback) | ISBN 9789811258992 (ebook for institutions) | ISBN 9789811259005 (ebook for individuals)
Subjects: LCSH: Bioinformatics.
Classification: LCC QH324.2 .P693 2023 | DDC 570.285--dc23/eng/20220713
LC record available at https://lccn.loc.gov/2022022559

British Library Cataloguing-in-Publication Data
A catalogue record for this book is available from the British Library.

For any available supplementary material, please visit
https://www.worldscientific.com/worldscibooks/10.1142/12908#t=suppl

Typeset by Stallion Press
Email: enquiries@stallionpress.com

Foreword from the First Edition

Olivo Miotto

Several years ago, when the first draft of the human genome was being completed, I decided to focus my efforts on the study of pathogen genomes. Armed with a background in software engineering, one of the first things that preoccupied me was a problem that loomed on the horizon and had little to do with the fascinating biology that was emerging from the study of genomes. It was already clear that, in order to study genetic variations, their effects on phenotype, and their epidemiological dynamics, it would be necessary to collect massive amounts of data, far more than most of us could actually handle. The question was not so much whether storage or processing capabilities would be sufficient — Moore's Law had accustomed us to rapid growth in computing power, and I was confident these technical challenges could be met. The critical question was whether the people who would be analysing these data would have sufficient know-how and resources to handle these large quantities of data, and extract the knowledge they needed. To be sure, the same problem was faced by companies that needed to build search engines, hotel booking systems, web-based ratings software, and all the services based on what we now call "big data". But genomics looked like a problem that could not be tackled by computer scientists alone. Biologists had to be empowered to handle scary amounts of data.

Those issues were evident even before whole-genome sequencing was revolutionized by the next-generation sequencing (NGS) technologies introduced by companies such as Solexa (now Illumina). Today, the MalariaGEN genomic epidemiology project on which I work (malariagen.net/projects/p-falciparum-community-project) comprises the genomes of *Plasmodium* parasites from almost ten thousand clinical samples, each backed by several gigabytes of short-read sequencing data — far more data than I would have predicted a few years ago. And yet, the knowledge gap has not been properly filled: if anything, it has become increasingly harder for life scientists and clinicians to effectively process such massive quantities of data, and many projects rely on collaborations with informatics specialists who often have limited expertise of the biological domain.

In the light of these difficulties, I give full credit to Lloyd Low and Martti Tammi for making a significant contribution towards filling the gap. What they have produced is a very practical guide, part reference and part tutorial, that will be appreciated by many life scientists for its direct and straightforward approach. Crucially, the content of this book is based on years of teaching experience, and "fine-tuned" by keeping in mind the difficulties routinely faced by those learning how to deal with NGS data. It contains a useful toolkit of techniques and practices using some of the most popular tools in use, such as BWA, samtools and so on.

The material covered in this book will support a broad range of applications: the final chapter suggests some possibilities, but clearly each reader will have to tackle challenges unique to their own areas of study, and this work will serve as a base on which to build further techniques. Commendably, it promotes the definition of a well-organized analytical workflow, and gives prominence to the quality aspects of genomics work — hugely important and frequently underestimated. Conducting a GWAS — or constructing a phylogeny — without first properly evaluating what data to rely upon and what to discard will invariably lead to useless or false results. It is therefore essential to instil high standards of quality into the mind of students and anyone undertaking genomic analyses.

I wish all readers all the best in their endeavours in this complex field, which I hope they will find rich in rewards.

Olivo Miotto

Mahidol-Oxford Tropical Medicine Research Unit, Bangkok, Thailand
Wellcome Trust Sanger Institute, Hinxton, United Kingdom
Centre for Genomics and Global Health, Oxford University, United Kingdom

Foreword from the First Edition

Nazar Zaki

The revolution that Next Generation Sequencing (NGS) brought to genetics can be compared to the revolution that the invention of the telescope brought to astronomy. Genetic phenomena can now be studied at the molecular level and genetic processes can be studied at genomic, transcriptomic and epigenomic levels using NGS technologies. The low cost of sequencing is allowing human genomes to be sequenced routinely and longer sequencing lengths allow the easier construction of novel genomes. Therefore, it is essential for researchers working in biology to have a good grasp of basic concepts in handling NGS data at different levels. This book provides a succinct and easy to read introduction to the processing of NGS data at various levels for a general audience.

For the novice user, the first three chapters provide a brief primer to the technology behind NGS and how to get past the hurdle of aligning NGS data to a reference genome. The alignment step is demonstrated using the popular open source aligner BWA and the commercial NovoAlign aligner that is known for its high accuracy. This chapter is written by an engineer at Novocraft itself and the reader can customize the workflow to achieve the required degree of precision and speed using NovoCraft products or open source options.

Once past the hurdle of aligning the reads, this book answers what naturally comes into mind: "What do I do next"? It introduces IGV so that the users can visualize the alignments and as the next step introduces the Galaxy framework to create a research workflow. Even if the user is not an expert in computer science, Galaxy will empower him to establish some basic research tasks after some experimenting. Overall, the reader can start diving deeper into analysing NGS data on his own after reading the first five chapters of the book.

While most of the NGS analysis currently starts with alignment, there are other applications that require genome assembly. This is especially true for smaller genomes and it is becoming popular as NGS technologies that produce very long read lengths are made available. In future it may be the case that the borderline between sequence alignment and assembly will not be clear cut. In Chapter 6, Dr. Tammi shares his expertise on sequence assembly with a gentle introduction to the basics of sequence assembly. Not only does he show the reader how to assemble a genome, but he also teaches how to gauge the quality of an assembly.

In the next few chapters, the book concentrates on specific application of NGS. The book has picked a timely set of applications that are being widely used and the user is guided step-by-step on how to process data for each application. Exome sequencing has become an important branch of NGS due to its cost considerations and the higher depth of coverage. We also have the ability to take snapshots of cells in action using transcriptome sequencing. Another different branch that is benefitted by NGS is metagenomics, which tries to find answers about the total genomic content of samples in contrast to the previous applications we discussed. Another important question is how to extract the relationships between genotypes and phenotypes. All these applications need different approaches and asks different types of questions. However, techniques used in these areas can be carried over to other methods. For example, techniques used for processing exome sequencing can be useful in working with other targeted sequencing methods and techniques used to find variations in WGS can be used in transcriptomic studies. Therefore, the reader can benefit by understanding the concepts used to process these different types of data sets.

Nazar Zaki

Professor

Leader, Bioinformatics Research Team

Coordinator, Intelligent Systems

Coordinator, Software Development

College of Information Technology

United Arab Emirates University

Al Ain 17551, UAE

Preface

The secret of life is encoded in DNA sequences. Since the 1970s, many inventors and innovators have enhanced DNA sequencing technologies to enable us to move from the painstaking process of reading a single base to now being able to easily gather the sequences of billions of DNA fragments. Today, we live in the era where next generation sequencing (NGS) technologies are commonly available and third generation sequencers have also found their niches in the market e.g. long read based genome assembly. NGS is well positioned to replace array-based SNP genotyping given advances in low pass whole genome sequencing, sequence capture technology and imputation techniques. Therefore, the trend for more NGS based applications continues to grow. New users of NGS usually have not worked with Sanger sequenced data and their introduction to this field is a straight jump into a dizzying amount of sequences. It is an understatement to say that it is difficult to handle the massive amount of sequenced data and to use them to make biological discoveries.

The idea for this book was conceived after my colleagues and I had organized and taught at various workshops on NGS. We thought that it would be a great idea to provide a comprehensive practical oriented book on NGS so that more people can learn how to handle bioinformatics data that are coming from this technology. The book covers general topics on how to handle NGS data from sequence quality inspection, alignment of reads to finding single nucleotide polymorphisms (SNPs). Other advanced topics such as genome assembly, exome sequencing, transcriptomics, metagenomics, genome wide association studies (GWAS) and machine learning are also covered. The chapter on GWAS is dedicated show the applications of NGS data to give readers a taste of the power of this technology in genetic mapping. The chapter on the use of machine learning methods to predict enhancers is another highlight of this book as it presents a gentle yet practical introduction to use machine learning to study important biological features.

There are common difficulties faced by many first time learners who need to analyze NGS data. This book put together materials and experiences gained from teaching many first time learners and it includes additional resources aimed at strengthening the readers knowledge in this field. We anticipate that this book will be of great use to students and researchers in the life sciences. For readers who are already proficient in NGS based data analysis, they can still keep the book as a reference material.

Note to readers: Companion datasets can be downloaded at https://github.com/PU-SDS/ngs-book-dataset

Contents

Acknowledgements

First and foremost, I must thank Joy Quek of the World Scientific Publishing for encouraging me to do a major update of my first book on next generation sequencing (NGS). After discussion with her and Martti, we thought we should cover machine learning, which has been useful in my own research on NGS data and seems popular to others. In addition, I am thankful to Dr. Asif Khan of Bezmialem Vakıf University and Perdana University for giving me the idea to write a book on NGS and bioinformatics. I also wish to thank Prof Bjorn Andersson of Karolinska institutet for spending some time to review the book. Last but not least, I wish to thank authors from the Sime Darby Technology Centre and the University of Adelaide for contributing book chapters. Without these key people, the book would not have been possible.

Lloyd Low

Chapter 1

Introduction to Next Generation Sequencing Technologies

Lloyd Low[a] and Martti T. Tammi[b]

[a]Perdana University Centre for Bioinformatics (PU-CBi), Block B and D1, MAEPS Building, MARDI Complex, Jalan MAEPS Perdana, 43400 Serdang, Selangor, Malaysia
[b]Biotechnology & Breeding Department, Sime Darby Plantation R&D Centre, Selangor, 43400, Malaysia

A Brief History of DNA Sequencing

In 1962, James Watson, Francis Crick, and Maurice Wilkins jointly received the Nobel Prize in Physiology/Medicine for their discoveries of the structure of deoxyribonucleic acid (DNA) and its significance for information transfer in living material.[1] The secret of DNA in orchestrating living activities lies in the arrangement of the four bases (i.e. adenine, thymine, guanine, and cytosine). The linear sequence of the four bases can be considered as the language of life with each word specified by a codon that is made up of three bases. It was an interesting puzzle to figure out how codons specify amino acids. In 1968, Robert W. Holley, HarGobind Khorana, and Marshall W. Nirenberg was awarded the Nobel Prize in Physiology/Medicine for solving the genetic code puzzle. Now it is known that the collection of codons direct what, where, when, and how much proteins should be made. Since the discovery of the structure of DNA and the genetic code, deciphering the meaning of DNA sequences has been an ongoing quest by many scientists to understand the intricacies of life.

The ability to read a DNA sequence is a prerequisite to decipher its meaning. Not surprisingly then, there has been intense competition to develop better tools to sequence DNA. In the 1970s, the first revolution in DNA sequencing technology began and there were two major competitors in this area. One was the commonly known Sanger sequencing method[2,3] and another was the Maxam–Gilbert sequencing method.[4] Over time, the popularity of the Sanger sequencing method and its modifications grew so much that it overshadowed other methods until perhaps 2005 when Next Generation Sequencing (NGS) began to take off.

In 1977, Sanger and colleagues successfully used their sequencing method to sequence the first DNA-based genome, a <X174 bacteriophage, which is approximately 5375 bp.[5] This discovery heralded the start of the genomics era. Initially, the Sanger sequencing method in 1975 used a two-phase DNA synthesis reaction.[2] In the first phase, a DNA polymerase was used to partially extend a primer bound onto a single-stranded DNA template to generate DNA fragments of random lengths. In phase two, the partially extended templates from the earlier reaction were split into four parallel DNA synthesis reactions where each reaction only had three of the four deoxyribonucleotide triphosphates (dNTPs; which are made up of dATP, dCTP, dGTP, dTTP). Due to a missing deoxyribonucleotide triphosphate (e.g. dATP), the DNA synthesis reaction would stop at its 3′ end position just one position prior to where the missing base was supposed to be incorporated. All of these synthesized DNA fragments could then be separated by size using electrophoresis on an acrylamide gel. The DNA sequence could be read off a radioautograph since its DNA synthesis happened with the incorporation of radiolabeled nucleotides (e.g. S-dATP).[35]

There were many problems with the initial version of the Sanger sequencing method that required further innovations before its widespread use and this scenario is akin to what is happening in the recent NGS technological developments. Some problems of the early Sanger sequencing method included the cumbersome two-phase procedures, only short length of a DNA sequence could be determined, the requirement of a primer meant some sequences of the template had to be known, hazardous radio labeled nucleotides were used and there was also no automated way to read off a DNA sequence. Sanger and colleagues rapidly improved on the method described in 1975 by eliminating the two-phase procedure with the use of dideoxynucleotides as chain terminators.[3] Briefly, the improved method started with four reaction mixtures that already had the single-stranded DNA template hybridized to a primer. In each reaction, the DNA synthesis proceeded with four deoxyribonucleotide triphosphates (one with radiolabeled nucleotide) and one dideoxynucleotide (ddNTP). Whenever a dideoxyribonucleotide was incorporated, the reaction terminated and thereby produced a mixture of truncated fragments of varying lengths. These DNA fragments were then separated by electrophoresis and then read off from a radioautograph. By adjusting the concentration of ddNTPs, chain termination can be manipulated to produce a longer sequence read.

To solve the requirement of knowing some template sequences for primer design, cloning was introduced. For example, the M13 sequencing vector is commonly used as a holder for DNA insert and known primers that bind to the vector sequence are available to be used to sequence the unknown DNA insert. One major innovation to the Sanger sequencing method is the replacement of radioactive labels with fluorescent dyes.[6] Four different dye color labels are available for the four dideoxynucleotide chain terminators and thus, DNA fragments that terminate at all four bases can be generated in a single reaction and thus analyzed on a single lane of acrylamide gel. The electrophoresis is coupled to a fluorescent detector that is also connected to a computer and thus sequence data can be automatically collected. In 1986, Applied Biosystems commercialized the first automated DNA sequencer (i.e. Model 370A) that is based on the Sanger sequencing method. For an animation of the Sanger sequencing method, the reader should refer to the Welcome Trust Sanger Institute (http://www.wellcome.ac.uk/Education-resources/Education-and-learning/Resources/Animation/WTDV026689.htm). Due to limitations of the chain terminator chemistry and resolution of the electrophoresis method, the Sanger sequencing method is only capable of sequencing a read of about 500–800 bases long. Most genes and other interesting DNA sequences are longer than that. Therefore, a method is required to first break up a longer DNA molecule into fragments, sequence the individual fragments and then piece them together to create a contiguous sequence (i.e. contig). In one approach known as the shotgun sequencing, the long DNA fragment is randomly sheared and then cloned for sequencing.[7] A computer program is then used to assemble the sequences by finding overlaps. It is challenging to find sequence overlaps when thousands to millions of DNA fragments are generated. The problem requires alignment algorithms and some notable examples of early work in this area include the Needleman–Wunsch algorithm[8] and Smith–Waterman algorithm.[9] Details on the bioinformatics involved in NGS alignment tools and sequence assembly are given in Chapters 4 and 6, respectively.

Next Generation Sequencing Technologies

One of the goals of the Human Genome Project (HGP) is to support advancements in DNA sequencing technology.[10] Although the HGP was completed with the Sanger sequencing method, many groups of researchers were already tinkering with new ideas to increase throughput and decrease cost of sequencing prior to the announcement of the first human genome draft in 2001. For example, developments for nanopore sequencing can be traced back to 1996 when researchers experimented with α-hemolysin.[11] After years of experimentations, the second DNA sequencing technology revolution finally took off in 2005 and ended Sanger sequencing dominance in the marketplace. The revolution is still ongoing at the time of this writing and it can be seen from the rapid decline in the cost of sequencing since the introduction of NGS technologies (Figure 1).

The sequencing technologies associated with the second revolution are referred to by various names, including second-generation sequencing, NGS, and high throughput sequencing. It should

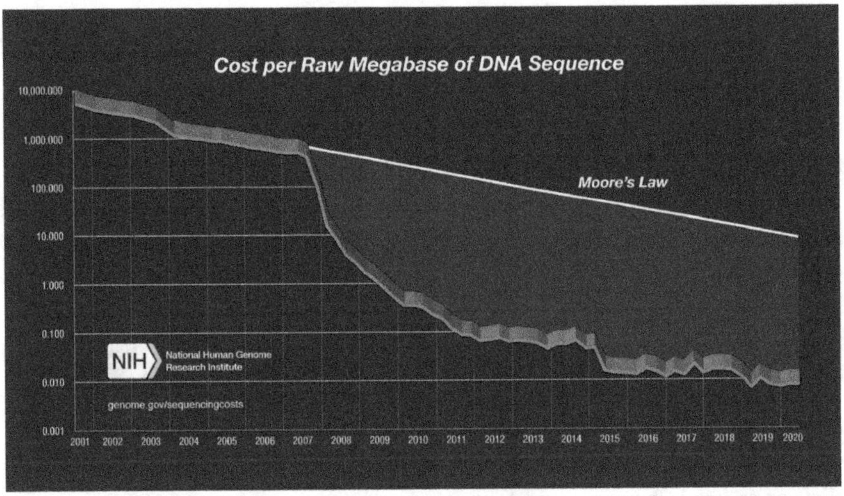

Figure 1. The cost to sequence one million bases of a specified quality (i.e. a minimum Phred score of Q_{20} for Sanger sequencing and an equivalent of Q_{20} or higher accuracy for NGS data) according to the National Human Genome Research Institute (NHGRI).[12] The cost of sequencing only made its rapid reduction in price from 2008 onwards.

perhaps be most appropriately termed as high throughput sequencing but NGS seems to be more commonly used to categorize such technologies and hence, this term is used for the book. For the purpose of this book, NGS technology refers to platforms that are able to sequence massive amount of DNA in parallel with a simultaneous sequence detection method and overall achieve a much cheaper cost per base than Sanger. These platforms include 454, ABI Supported Oligo Ligation Detection (SOLiD), Illumina, and Ion Torrent. Due to the popularity of the Illumina platform at the time of this writing, the practical chapters (i.e. Chapters 3–10) of the book emphasize on the use of Illumina data as sample datasets.

There is a third revolution in sequencing technology underway with the commercialization of third-generation sequencing technologies such as those from Pacific Biosciences and Oxford Nanopore Technologies. Third-generation sequencing is defined as the sequencing of single DNA molecules without the need to halt between read steps, whether enzymatic or otherwise.[13] There are three categories of single-molecule sequencing: (i) sequencing by synthesis method whereby base detection occur real-time (e.g. PacBio), (ii) nanopore technologies whereby DNA thread through a nanopore and are detected as they pass through it (e.g. Oxford Nanopore), and (iii) direct imaging of DNA molecules using advanced microscopy (e.g. Halcyon Molecular (this company has shut down)).

DNA sequence data generation process among different sequencing platforms may share similarities such as the general "wash and scan" approach but they may differ in terms of cost, runtime, and detection methods. The sequence data from different platforms have different characteristics such as error patterns and different tools being used to process the raw data to FASTQ format. Much of the internal workings of NGS sequencers are proprietary matters and users generally rely on providers to come out with their own tools for base calls as well as error calls. After that, a sequence is assumed as "correct" and researchers proceed to analyze it. The subsequent sections aim to introduce the background and some details of commercially available platforms, which include 454, ABI SOLiD, Illumina, Ion Torrent, PacBio, and Oxford Nanopore. Besides these six platforms, there are other companies out there that also innovate in this space such as SeqLL, GnuBIO, Complete Genomics, and others, but they will not be covered here. For a list of available sequencing companies, readers are encouraged to read a news article by Michael Eisenstein in 2012 that was published by *Nature Biotechnology*, which detailed 14 NGS companies.[14]

454

A company named 454 Life Sciences Corporation made the first move in the NGS revolution. The company was initially majority owned by CuraGen. It was from this company that the name "454" originated, which was just a code name for a project. 454 was later acquired by Roche in 2007. It made a public announcement in 2003 that it managed to sequence the entire genome of a virus in a single day.[15] Then in 2005, scientists using 454 technology published an article in *Nature* on the complete sequencing and *de novo* assembly of *Mycoplasma genitalium* genome with 96% coverage and 99.96% accuracy in one run of the machine.[16] In the same year, the company made a system named Genome Sequencer 20 (GS20) commercially available. This breakthrough in sequencing throughput and speed was an incredible feat when compared to the Sanger technology and it created a lot of excitement.

The principle behind 454 relies on pyrosequencing, which was a technology licensed from Pyrosequencing AB. This method depends on the generation of inorganic pyrophosphate (PPi) during PCR when a complementary base is incorporated[17] (Figure 2).

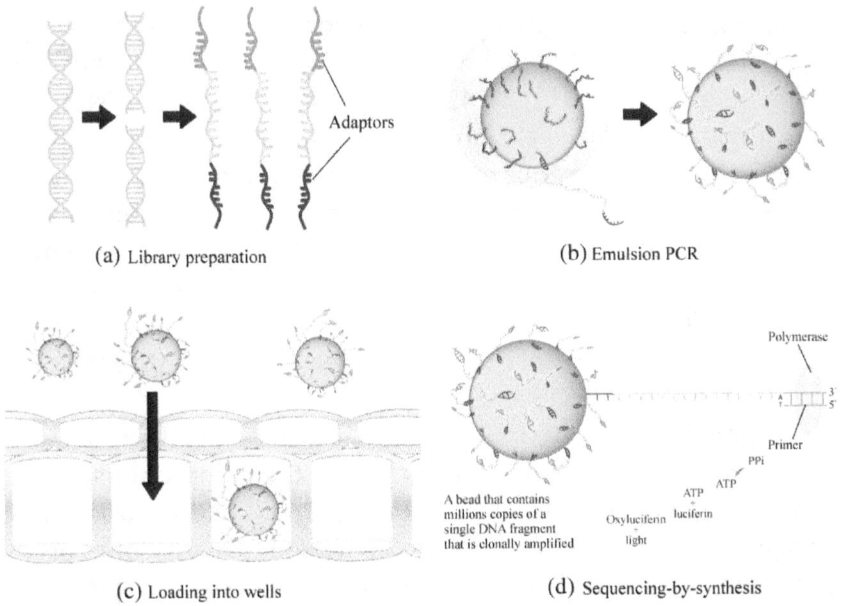

Figure 2. 454 pyrosequencing method. (a) In brief, the method starts with a single-stranded library that has adaptors on both ends. (b) The adaptor sequence is used to bind to the bead. This is followed by emulsion PCR to generate millions of copies of single DNA fragment on each bead. (c) After that, beads are placed into a device known as PicoTiter Plate for sequencing by detection of base incorporation during PCR. (d) Whenever a base is incorporated, inorganic pyrophosphate (PPi) is generated. PPi is converted to ATP by sulfurylase and luciferase uses the ATP to convert luciferin to oxyluciferin and light.

PPi is converted to ATP by sulfurylase and luciferase uses the ATP to convert luciferin to oxyluciferin and light. The reaction occurs very fast, in the range of milliseconds, and the light produced can be detected by a charge-couple-device (CCD) camera. One of the key innovations of 454 technology is miniaturization of the pyrosequencing reactions, thereby allowing for parallel sequencing reactions to occur in a small space using smaller volume of reagents. Another innovation is simultaneous detection of the light signals from many individual reactions.

One of the key drawbacks of the 454 pyrosequencing chemistry is the difficulty in detection of the actual number of bases in homopolymer tract (e.g. AAAAA). There is no blocking mechanism included to prevent multiple same bases incorporation during DNA elongation and thus light signals are stronger in longer homopolymer tracts. The light signal is actually light intensity that is converted to a flow value in the 454 system. It is difficult to distinguish how many bases there are once the homopolymer is more than 8 bases long.[16] The presence of homopolymers is the reason why 454 sequence reads do not have fixed lengths, unlike the Illumina platform that includes a blocking mechanism that allow the reading of only a single base each time. Another shortcoming of the 454 system is artificial amplification of replicates of sequences during the PCR step. It was estimated in a metagenomics study that this type of error is between 11% and 35%.[18]

Although a pioneer in NGS, 454 has officially lost the race of the sequencing game. As seen in Figure 3, on the comparisons of NGS platforms, the trend for 454 sequencing in articles tracked by Google Scholar has reached a plateau. It used to hold a lot of promises in revolutionizing sequencing and it was even regarded by some as the technology that had won the sequencing race. Roche announced the closing down of 454 in 2013.[19] Sequencers from 454 started being phased out in the middle of 2016.

ABI SOLiD

The initial success story of 454 sequencers challenged the dominance of Applied Biosystems (AB), which was the main supplier of

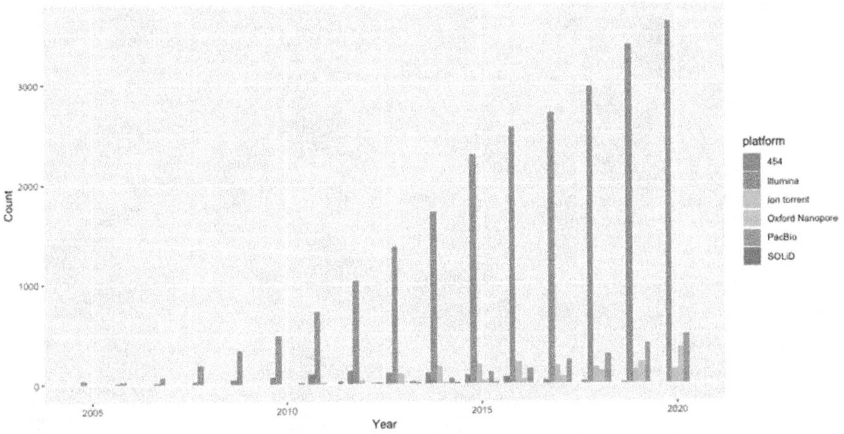

Figure 3. Comparisons of popularity of NGS platforms over the years by using keywords as search terms in PubMed. The keywords for searches are as follow: 454 — "454 Sequencing"; Illumina — "Illumina"; PacBio — "Pacbio"; SOLiD — "SOLiD sequencing"; Ion torrent — "Ion torrent"; Oxford nanopore — "Oxford nanopore".

Sanger-based sequencing machines for the HGP. The ABI PRISM 3700 was a very popular system and many researchers who needed to perform sequencing prior to 2005 were familiar with the system. In 2006, ABI completed acquisition of Agencourt Personal Genomics, which allowed it to market a novel NGS technology known as SOLiD. Currently, Thermo Fisher Scientific owns SOLiD sequencing technology after it acquired Life Technologies, which is a company formed from the merging of Invitrogen and AB. From Figure 3, it seems that SOLiD sequencing is not that popular as a NGS platform when compared to the others even though it has been available since 2006. To our knowledge, SOLiD is the only NGS platform that employs ligation-based chemistry with a unique di-base fluorescent probes system.

Understanding the SOLiD sequencing system is akin to solving a jigsaw puzzle due to the di-base encoding system. The sample preparation steps prior to probes ligation are very similar in concept to the 454 system. Briefly, a genomic DNA library is sheared into smaller fragments and both ends of each fragment will be tagged with different adaptors (e.g. Adaptor P1 — Fragment 1 — Adaptor P2).

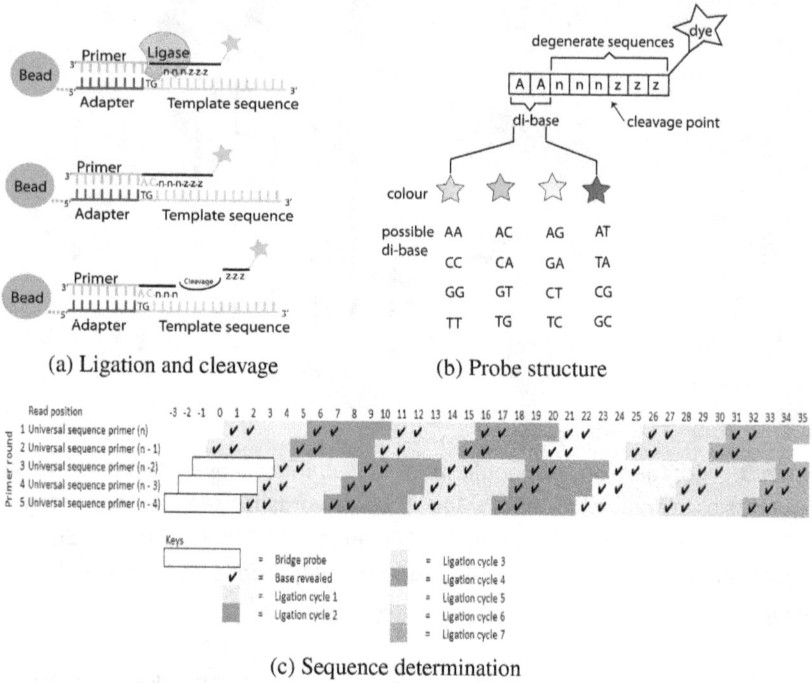

Figure 4. An overview of the SOLiD sequencing process. (a) Each ligation cycle starts with the 8-mer probe binding to the template and then ligated for its detection. Then, cleavage occurs to remove three nucleotides and a tagged dye. (b) The structure of the 8-mer probe. (c) An illustration of the sequence determination process during each ligation cycle of the primer rounds. Position 0 is a part of the adaptor sequence and template sequence is only revealed from position 1 onwards.

Then emulsion PCR will take place to create beads enriched with copies of the same DNA fragment on each bead. The beads are then attached to a glass slide through covalent bonds. From here ligation and detection of bases will take place (Figure 4(a)). Firstly, a universal sequence primer (*n*) is used to bind to the known adaptor sequence. Then a specific 8-mer probe with sequence-structure as depicted in Figure 4(b) will out compete other probes for binding immediately after the primer-binding site. Ligation then occurs and identity of the bound probe is detected by distinguishing which fluorescent dye is tagged at the probe's 5′ end. Then cleavage occurs at a position between the fifth and sixth nucleotide of the probe. After cleavage is complete, subsequent ligation is possible as a free phosphate group is now available at the fifth base of the probe. The reason why only one particular 8-mer probe will win the binding site is due to the specific di-base sequence at the 3′ end that distinguishes the collection of probes. Only four types of fluorescence dyes are used and each 8-mer probe with specific di-base sequence is tagged by a dye at the 5′ end. This system is unique in the sense that a di-base sequence is detected in each ligation cycle.

The ligation and cleavage process can be repeated many times to achieve the desired sequence length. However, it will only give sequence information two bases at a time with a gap of 3 bases in between. Next, the ligate-cleave-detect process is repeated with a new universal primer (*n*–1), which is a primer that binds exactly one base further upstream at the 5′ end of the adaptor sequence. This ligate-cleave-detect process that cycles a few times with a new primer is also known as reset. The entire process is repeated another three more rounds with universal primer (*n*–2), (*n*–3), and (*n*–4). Altogether, five different universal primers are used. Figure 4(c) shows an example of sequence determination after five rounds of reset. Note that each base is called twice in independent primer round and this increases the accuracy of base call. A check for concordance of the two calls for the same base represents an in-built error checking property of this system and allows it to achieve an overall accuracy greater than 99.94%. Although the SOLiD system is unique in the sense that it can store sequence of oligo color calls (i.e. color space) to be used for mutation calls, this method does introduce challenges to bioinformatics analysis as most tools are based on DNA calls rather than color space model.

Illumina

In the mid-90s, Shankar Balasubramanian and David Klenerman, both from the University of Cambridge conceived the idea of massive parallel sequencing of short reads on solid phase using reversible terminators. They formed Solexa in 1998 after successfully received funding from a venture capital firm. The sequencing approach by Solexa is also known as sequencing-by-synthesis. The company launched its first sequencer, Genome Analyzer in 2006 and the machine is capable of producing 1 Gb of data in just a single run. Figure 5 shows an overview of the Illumina sequencing-by-synthesis method.

Illumina acquired Solexa in 2007. Soon after its acquisition, there were at least three high profile research publications in Nature 2008 volume 456, which highlighted the capabilities of the Genome Analyzer in sequencing human genomes (e.g. African

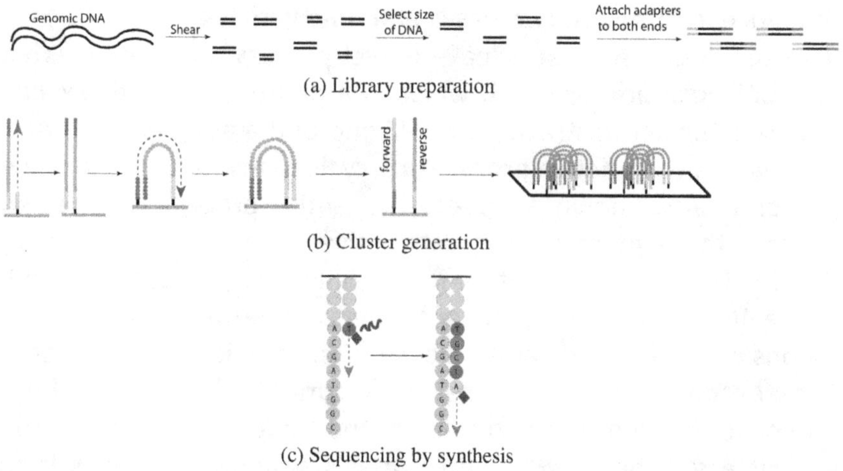

Figure 5. An overview of the Illumina sequencing process. (a) Genomic DNA is sheared, size selected, and then attached with adaptors at both ends. (b) The DNA library is placed on the flow cell to allow for complementary binding at one end of the adaptor to probes that are coated on the surface. Then bridge amplification in the solid phase occurs to generate clusters of single DNA fragments. After that, reverse strands are cleaved and washed away. A cluster of clonal sequences is required to enable a high signal to noise during base detection. (c) Sequencing begins with a primer binding to the remaining forward strand and a DNA polymerase is used to incorporate the right fluorescently labeled nucleotide among the four possible options (i.e. A, C, T, or G). At each cycle, only one nucleotide is incorporated due to the use of reversible terminator chemistry and detection occurs at this stage. This is followed by a cleavage step and the next cycle is ready to go.

genome,[20] Chinese genome,[21] and cancer patient genome[22]). In the subsequent years, the popularity of this system grew so much that by 2020, the cumulative number of articles that cited Illumina was over twenty thousand (Figure 3). To quote a marketing brochure by Illumina in 2015, "More than 90% of the world's sequencing data is generated using Illumina sequencing-by-synthesis method." The company is also very creative at developing and marketing their products with sequencing systems (e.g. MiniSeq, MiSeq, MiSeqDx, NextSeq 500, HiSeq 2500, HiSeq 3000, HiSeq 4000, HiSeq X Ten, HiSeq X Five) that suit researchers who operate on different budgets and require different level of sequencing throughput. The Illumina systems can be used for a wide range of applications that include resequencing, whole-genome sequencing, exome sequencing, metagenomics, epigenetic studies, and sequencing of a panel of genes such as targeting genes linked to cancer (e.g. TruSight Cancer).

One of the key strengths of the Illumina platform is the ability to produce high throughput of DNA sequence data at a lower cost despite only producing short sequences (e.g. paired-end of 35 bp in the African genome sequencing[20]). Improvement in bioinformatics methods allows researchers to do so much more than what was thought as possible if only short, accurate reads are available. Nowadays, the Illumina system can produce paired-end sequences of 300 bp for each end, which further enhances the power of this technology. Besides the advantage of high throughput low-cost sequencing, it also performs better than the 454 system with respect to homopolymer sequencing error because it uses reversible terminator sequencing chemistry. Only a single base is incorporated each time prior to detection in the Illumina system whereas 454 allows for multiple bases incorporation in a homopolymer tract. However, the Illumina system also comes with drawbacks. The 3′ end of the sequence tends to be of lower quality than the 5′ end, which means some sequences from the 3′ end should be filtered out if it is below certain set threshold (see Chapter 3). There can also be tiles associated error when the flow cell is affected by bubbles in reagents or some other unknown causes.[23] In addition, sequence-specific errors have also been found for inverted repeats and GGC sequences.[24] Furthermore, in a study on 16S rRNA amplicon sequencing on the MiSeq, library preparation method and choice of primers significantly influence the error patterns.[25]

Ion Torrent

Besides SOLiD sequencing, Thermo Fisher Scientific has another NGS platform on its portfolio known as Ion Torrent, which was acquired from Life Technologies. Initially, Life Technologies developed the platform and released the Ion Personal Genome Machine (PGM) in 2010. The launch of this machine created much excitement among researchers who wanted affordable sequencers for their laboratories. It was sold at just $49,500 per sequencer and utilized cheap disposable chip of about $250.[26] In addition, it runs faster when compared with competing machines such as HiSeq from Illumina. However, in terms of DNA data throughput, it loses out in comparison to the Illumina HiSeq.

Like the 454 and SOLiD systems, the library preparation and emulsion PCR steps on beads are present in the Ion Torrent. The main difference lies in the detection of nucleotide incorporation that is not based on fluorescence or chemiluminescence, but instead it measures the H^+ ions released during the process. In other words, detection of nucleotide incorporation is done by miniature semiconductor pH sensor. Since each of the four DNA bases are supplied sequentially for DNA elongation, if the base matches the template, then a signal is detected. For homopolymer region in the template, the signal will be amplified but accurate detection on the actual number of bases is challenging.[27] Only natural nucleotides are needed and no high-resolution camera and complicated image processing are required, which when taken together are some reasons for a faster runtime and lower machine cost. For a video on the Ion Torrent method, the reader should refer to the Thermo Fisher Scientific sequencing education webpage (http://www.thermofisher.com/my/en/home/life-science/sequencing/sequencing-education.html#).

Following the release of Ion PGM, the Ion Torrent product line now includes Ion Proton, Ion Chef, and Ion S5 systems. There is a diverse range of applications for these systems such as targeted sequencing, exome sequencing, transcriptome sequencing, bacterial and viral typing. However, genomic studies that involve *de novo* assemblies of larger genomes (e.g. >1 Gbp) do not seem to be the target areas of Ion Torrent. The popularity of Ion Torrent has been steadily rising but seemed to have reached a plateau (Figure 3).

Pacific Biosciences

The second-generation sequencing technologies are generally characterized by the "wash and scan" procedure that is much slower than the natural rate of DNA elongation by DNA polymerase. Furthermore, the length of contiguous DNA that can be sequenced is rather short (e.g. <1 kb). If one could observe DNA polymerization in real-time and detect which base was incorporated each time there was a DNA polymerase activity, faster sequencing runtime, and longer read length could be achieved. However, there are many challenges for detection of bases incorporation during real-time DNA polymerase activities because they happen too fast.

Pacific Biosciences, which was founded in 2004, has made two key innovations that enabled real-time observation of DNA synthesis.[28] One of them is the use of phospholinked nucleotides. Each phospholinked nucleotide has a fluorescent dye attached to the phosphate chain rather than to the base. During DNA elongation, the phosphate chain is cleaved and hence the dye label diffuses away. The DNA template is ready to accept the next nucleotide. Another key innovation is the use of zero-mode waveguide (ZMW) as the platform for detection of base incorporation. These ZMWs are housed inside a SMRT Cell. A ZMW can be thought of as a well with a very tiny hole at the bottom, which enables visible laser light to pass through. However, the light intensity decays exponentially and thus it can only illuminate the bottom of the well. With a DNA polymerase immobilized at the bottom of the well, its DNA polymerase activity can be monitored as it is illuminated. This is akin to having a miniature microscope placed at the bottom to peek at DNA polymerase activity on top of it. Phospholinked nucleotides diffuse into the well and when the right one is encountered by the DNA polymerase, it will be incorporated to the growing strand. The simple diffusion of phospholinked nucleotides happens in the microseconds range but when they are incorporated to the growing DNA strand, they stay longer at the site of incorporation (i.e. miliseconds range). It is from this longer stay by a particular phospholinked nucleotide that a signal is detected against a background of other free moving nucleotides.

Another interesting aspect of the PacBio technology is the observation of the kinetics of DNA polymerase activity. Kinetics data can be collected directly from the system and this allows for an investigation of favorable mutations of DNA polymerase with a lower sequencing error rate. In addition, environmental parameters such as pH, temperature, and concentration of inhibitor that affect the kinetics of DNA polymerase can also be optimized. For researchers interested in epigenetics, the PacBio system is able to detect epigenetic effects such as base methylation (e.g. N^6-methyladenine (m^6A) and N^4-methylcytosine (m^4C)) because such modification to the DNA template affects the kinetics of DNA polymerase.

A detailed report on the PacBio technology was first published in Science in 2009.[28] The company released their commercial platform PacBio RS in 2011 and later the PacBio RS II in 2013. It is rather impressive that the combination of PacBio RS II with P6-C4 chemistry can achieve an average read length of 10–15 kb. The combination of an upgrade in the PacBio machine to the higher throughput Sequel and newer sequencing chemistry v3 have enabled read length to an average of 30 kb polymerase read length. The library size for SMRT sequencing ranges from 250 bp to 50 kbp. As the main advantage of the PacBio system is its long read length, researchers have tried to use its sequenced data alone or in combination with other sequenced data to *de novo* assemble various genomes including bacteria (e.g. *Escherichia coli*), yeast (e.g.

Saccharomyces cerevisiae), plant (e.g. *Arabidopsis thaliana*) and animals (*D. melanogaster, Homo sapiens*).[29] It is now known that PacBio technology is particularly good for closing gaps in *de novo* assembled genomes, resolve phases among haplotypes, produce full-length RNA transcripts isoforms sequences, identify structural variants and to sequence complex regions with repeats.[30] However, its main disadvantages are its low throughput, high cost per sequenced base and high error rate (~ 11–15%). The errors are not biased towards homopolymers but appear as random with indels errors more common than substitution errors. Owing to the random error feature, if there is enough PacBio sequenced data coverage on a particular template, the consensus sequence can achieve a much higher accuracy than a single sequence pass. In late 2015, PacBio announced the release of the Sequel System that has a redesigned SMRT Cell, which now contains 1 million ZMWs. It provides 7× higher sequencing throughput than PacBio RS II and this development is exciting in terms of highlighting the scalability of this technology. Since then, PacBio has upgraded the Sequel machine to handle 8 million ZMWs. Additionally, the read accuracy of this platform has substantially improved with the introduction of HiFi reads (https://www.pacb.com/smrt-science/smrt-sequencing/hifi-reads-for-highly-accurate-long-read-sequencing/). For more information on the PacBio system, readers should refer to the company's website: http://www.pacb.com.

Oxford Nanopore Technologies

Besides PacBio, there is another new entrant to the sequencing race that also belongs to the third generation sequencing category — Oxford Nanopore Technologies. The company was a spin-off from the University of Oxford in 2005 and its goal is to democratize sequencing by making it affordable and portable (https://nanoporetech.com). The company's sequencers made its debut in 2012 at the Advances in Genome Biology and Technology meeting.[31] The sequencer MinION was introduced during the meeting but it was only in 2014 that a limited number of participants who were a part of MinION Access Programme (MAP) received their first sequencers for performance testing. Then in 2015, the first nanopore sensing conference known as the London Calling was held and researchers gathered to find out more about the MinION technology. In that same year, MinION was made commercially available. At the time of this writing, the company also has two other systems in development, PromethION, and GridION. Although new, the technology has occupied a rather interesting niche where portability of DNA sequencers is required, such as in real-time genomic surveillance of Ebola outbreak[32] and DNA sequencing in space to monitor changes to microbes and humans in spaceflight, as well as other astrobiological applications.[33]

The methodology behind the MinION technology was described in a whole-genome shotgun sequencing of a reference *Escherichia coli* strain.[34] The DNA library preparation method was elaborated in the mentioned paper. An ideal DNA fragment for sequencing has a DNA hairpin loop that is ligated on one end to join the two strands together. Then, one of the strands will traverse into a protein nanopore that is anchored on an electrically resistant polymer membrane. The setup of the nanopore is such that any analyte that passes through it or approach its opening will create a disruption in current. Measurements of the characteristics of this disruption then lead to identification of which nucleotides have passed through the pore. After the first strand has moved in, the other strand will follow suit. Similar to the PacBio, it is also possible to identify epigenetic modifications to the DNA using this method. The sequencing process is scalable by using more nanopores for simultaneous detection of DNA fragments that are moving through them.

The procedures involved sound simple and allow for the sequencing of a single long DNA molecule without amplification and usage of fluorescent dyes that require expensive imaging. This is clearly a case of a disruptive technology in the making but the technology is still characterized by high sequencing errors. In a paper that compared sequencing errors, the error rate of Oxford Nanopore technology is in the range of 20–25%.[35] More time is needed for the technology to mature and to improve on the error rate. The Oxford Nanopore sequencing delivers the longest sequencing read, ~2.3 Mb,[36] and it usually takes library size that ranges from 10 kb to 30 kb.

Informatics Challenges

Advances in sequencing technologies have enabled the scientific community to decode more than 65,000 organisms' genomes.[37] The trend for more sequenced data is likely to continue unabated. According to Raymond McCauley of the Singularity University, "It turns out that one human genome wasn't worth much, but thousands upon thousands represent an invaluable pool of data to be sifted for patterns and correlated with diseases, treatments, and outcomes."[38] To sift through massive amount of sequenced data is a challenge and to begin to address this problem, we need to increase the supply of skilled bioinformaticians. This is in fact one of the main reasons for writing this book. For beginners who need to use second or third-generation sequencing technologies, they will likely face informatics challenges in terms of knowing how each sequencer produces its raw sequence output, conversion of sequenced data to FASTQ format, quality checking, alignment to a reference, or *de novo* assembly, and interpretation of results (e.g. impact of SNPs, indels, etc.). Therefore, the subsequent chapters from here will focus on developing skills needed to navigate seas of NGS data in order to help answer biological questions.

References

1. Watson, J. D. & Crick, F. H. Molecular structure of nucleic acids; a structure for deoxyribose nucleic acid. *Nature* **171**, 737–738 (1953).
2. Sanger, F. & Coulson, A. R. A rapid method for determining sequences in DNA by primed synthesis with DNA polymerase. *Journal of Molecular Biology* **94**, 441–448 (1975).
3. Sanger, F., Nicklen, S. & Coulson, A. R. DNA sequencing with chain-terminating inhibitors. *Proceedings of the National Academy of Sciences of the U.S.A.* **74**, 5463–5467 (1977).

4. Maxam, A. M. & Gilbert, W. A new method for sequencing DNA. *Proceedings of the National Academy of Sciences of the U.S.A.* **74**, 560–564 (1977).
5. Sanger, F. *et al.* Nucleotide sequence of bacteriophage phi X174 DNA. *Nature* **265**, 687–695 (1977).
6. Smith, L. M. *et al.* Fluorescence detection in automated DNA sequence analysis. *Nature* **321**, 674–679 (1986).
7. Anderson, S. Shotgun DNA sequencing using cloned DNase I-generated fragments. *Nucleic Acids Research* **9**, 3015 (1981).
8. Needleman, S. B. & Wunsch, C. D. A general method applicable to the search for similarities in the amino acid sequence of two proteins. *Journal of Molecular Biology* **48**, 443–453 (1970).
9. Smith, T. F. & Waterman, M. S. Identification of common molecular subsequences. *Journal of Molecular Biology* **147**, 195–197 (1981).
10. Collins, F. S. New goals for the U.S. Human Genome Project: 1998–2003. *Science (80-).* **282**, 682–689 (1998).
11. Kasianowicz, J. J., Brandin, E., Branton, D. & Deamer, D. W. Characterization of individual polynucleotide molecules using a membrane channel. *Proceedings of the National Academy of Sciences of the U.S.A.* **93**, 13770–13773 (1996).
12. Wetterstrand, K. A. *DNA sequencing costs: Data.* <https://www.genome.gov/about-genomics/fact-sheets/DNA-Sequencing-Costs-Data>.
13. Schadt, E. E., Turner, S. & Kasarskis, A. A window into third-generation sequencing. *Human Molecular Genetics* **19**, R227–240 (2010).
14. Eisenstein, M. The battle for sequencing supremacy. *Nature Biotechnology* **30**, 1023–1026 (2012).
15. Pollack, A. Company says it mapped genes of virus in one day. *The New York Times* (2003).
16. Margulies, M. *et al.* Genome sequencing in microfabricated high-density picolitre reactors. *Nature* **437**, 376–380 (2005).
17. Royo, J. L. & Galán, J. J. Pyrosequencing for SNP genotyping. *Methods in Molecular Biology* **578**, 123–133 (2009).
18. Gomez-Alvarez, V., Teal, T. K. & Schmidt, T. M. Systematic artifacts in metagenomes from complex microbial communities. *ISME Journal* **3**, 1314–1317 (2009).
19. Roche Shutting Down 454 Sequencing Business. *Genomeweb.* <https://www.genomeweb.com/sequencing/roche-shutting-down-454- sequencing-business> (2013).

20. Bentley, D. R. *et al.* Accurate whole human genome sequencing using reversible terminator chemistry. *Nature* **456**, 53–59 (2008).
21. Wang, J. *et al.* The diploid genome sequence of an Asian individual. *Nature* **456**, 60–65 (2008).
22. Ley, T. J. *et al.* DNA sequencing of a cytogenetically normal acute myeloid leukaemia genome. *Nature* **456**, 66–72 (2008).
23. Dolan, P. C. & Denver, D. R. TileQC: a system for tile-based quality control of Solexa data. *BMC Bioinformatics* **9**, 250 (2008).
24. Nakamura, K. *et al.* Sequence-specific error profile of Illumina sequencers. *Nucleic Acids Research* **39**, e90 (2011).
25. Schirmer, M. *et al.* Insight into biases and sequencing errors for amplicon sequencing with the Illumina MiSeq platform. *Nucleic Acids Research* **gku**, 1341 (2015) **43**(6), 37(2015) . doi:10.1093/nar/gku1341.
26. Katsnelson, A. DNA sequencing for the masses. *Nature* (2010). doi:10.1038/news.2010.674.
27. Bragg, L. M., Stone, G., Butler, M. K., Hugenholtz, P. & Tyson, G. W. Shining a light on dark sequencing: Characterising errors in Ion Torrent PGM data. *PLoS Computational Biology* **9**, e1003031 (2013).
28. Eid, J. *et al.* Real-time DNA sequencing from single polymerase molecules. *Science* **323**, 133–138 (2009).
29. Berlin, K. *et al.* Assembling large genomes with single-molecule sequencing and locality-sensitive hashing. *Nature Biotechnol*ogy **33**, 623–630 (2015).
30. Rhoads, A. & Au, K. F. PacBio sequencing and its applications. *Genomics. Proteomics Bioinformatics* **13**, 278–289 (2015).
31. Check Hayden, E. Nanopore genome sequencer makes its debut. *Nature* doi:10.1038/nature.2012.10051 (2012).
32. Quick, J. *et al.* Real-time, portable genome sequencing for Ebola surveillance. *Nature* **530**, 228–232 (2016).
33. Rainey, K. *Sequencing DNA in the palm of your hand.* <http://www.nasa.gov/mission_pages/station/research/news/biomolecule_sequencer> (2015).
34. Quick, J., Quinlan, A. R. & Loman, N. J. A reference bacterial genome dataset generated on the MinION(TM) portable single-molecule nanopore sequencer. *Gigascience* **3**, 22 (2014).
35. Laehnemann, D., Borkhardt, A. & McHardy, A. C. Denoising DNA deep sequencing data-high-throughput sequencing errors and their correction. *Briefings in Bioinformatics.* bbv029-(2015). doi:10.1093/bib/bbv029.

36. Payne, A. *et al.* BulkVis: a graphical viewer for Oxford nanopore bulk FAST5 files. *Bioinformatics* **35**(13), 2193–2198 (2019).
37. Reddy, T. B. K. *et al.* The Genomes OnLine Database (GOLD) v.5: a metadata management system based on a four level (meta)genome project classification. *Nucleic Acids Research* **43**, D1099–106 (2015).
38. Greenwald, T. *DNA Tsunami: Raymond McCauley explains why bioinformatics is good for business — Forbes.* <http://www.forbes.com/sites/tedgreenwald/2011/10/20/dna-tsunami-raymond-mccauley-explains-why-bioinformatics-is-good-for-business/#555a2c1d2986>.

Chapter 2

Primer on Linux

Adeel Malik and Muhammad Farhan Sjaugi

Perdana University Centre for Bioinformatics (PU-CBi), Block B and D1, MAEPS Building, MARDI Complex, Jalan MAEPS Perdana, 43400 Serdang, Selangor, Malaysia.

Introduction

Many developers of NGS tools prefer to use Linux as the operating system for their works. To use these tools (e.g. BWA, Bowtie, and SAMTOOLS) users need to have a good level of proficiency in Linux. However, to our knowledge, most biologists who need to work with NGS are unfamiliar with the operating system and require at least a gentle introduction on this topic for them to better understand commonly used commands in Linux. Otherwise, they need to juggle with two difficulties while learning NGS tools; (i) the general Linux features and (ii) the new tools that they need to master. The aim of this chapter is to remove the first difficulty associated with familiarizing oneself with the Linux system so that users can concentrate on understanding NGS tools. It is not possible to cover all aspects of Linux but the intention here is for users to be able to navigate the rest of the chapters with ease. For users who are already familiar with Linux, they may skip this chapter and go directly to Chapter 3 on sequence quality.

Listing the Contents of a Directory

'ls' may probably be the first command that you use at the command prompt and the purpose is to list the contents of any directory. For example, when you log in to a Linux terminal and would like to list directories and files that are in your home directory (in this case, ngsguide), you will simply type 'ls' and then press the 'enter' key. Note that the term directories and folders are used interchangeably.

```
$ ls
```

In this example of 'ls' command, nothing is listed after pressing the 'enter' key from the keyboard. It is because in our case the current home directory (ngsguide) is empty and does not return anything on the screen. However, this may differ from system to system where you may have preexisting folders on your machine.

'ls' can also be used with a variety of available options. These additional options can be used individually with 'ls' or in combination as explained below.

The first option that we will use with 'ls' is '-l' (Please note: 'l' is L in lower case.)

```
$ ls-l
```

```
$ ls -l
total 0
```

When the 'ls -l' command is used, additional information about each file and directory such as size, file or folder name, owner of these files and their permissions, modified date and time, etc. is displayed. However, '0' is returned in our case. The reason is because there are no files or directories in the home directory.

'ls' in combination with '-a' can be used to list the hidden files or folders as follows.

```
$ ls-a
```

```
$ ls-a
. .. .bash_history .bash_logout .bash_profile
.bashrc .kshrc .mozilla .viminfo
```

Each hidden file or directory starts with a DOT character. Therefore, files such as .bash_history, .bashrc, .kshrc are hidden files, whereas .mozilla is a hidden folder.

Now let us try to combine '-l' and '-a' options with 'ls' command.

```
$ ls-la
```

```
$ ls -la
total 36
drwx------    3  ngsguide ngsguide 4096 Feb 18   03:11  .
drwxr-xr-x    44  root              root 4096  Feb 16  13:29  ..
-rw-------    1  ngsguide ngsguide  198    Feb    17  08:54  .bash_history
-rw-r--r--    1  ngsguide ngsguide  18     Jul    22  2015   .bash_logout
-rw-r--r--    1  ngsguide ngsguide  176    Jul    22  2015   .bash_profile
-rw-r--r--    1  ngsguide ngsguide  124    Jul    22  2015   .bashrc
-rw-r--r--    1  ngsguide ngsguide  171    Jul    22  2015   .kshrc
drwxr-xr-x    4  ngsguide ngsguide   4096  Apr    21  2014  .mozilla
-rw-------    1  ngsguide ngsguide  603    Feb    18  02:49  .viminfo
```

You can now see that by using 'ls -la' additional details about these hidden files and folders (or any other files or folders) can be obtained. Similarly, you can use other options available for 'ls' individually or in combination. To explain each parameter in detail is beyond the scope of this chapter. There are plenty of online resources available describing these commands and parameters in greater detail (for example: http://www.yourownlinux.com/2014/01/linux-ls-command-tutorial-with-examples.html; http://www.computerhope.com/unix/uls.htm).

In case you are not connected to the internet, a very handy Linux utility/command 'man' can be used to get help on any Linux command. For example, to get help on 'ls' command, just type the following command:

```
$ man ls
```

Using 'man ls' gives a detailed help on all the options available for 'ls' command. 'man' is the Linux system's guidebook and can be used to display manual pages for specific Linux commands (http://www.computerhope.com/unix/uman.htm). In this case, we asked Linux to display the help on 'ls' command as shown in the box below (sample output).

```
LS(1)                          User Commands                          LS(1)
NAME
  ls - list directory contents
SYNOPSIS
  ls [OPTION]... [FILE]...
DESCRIPTION
 List information about the FILEs (the current directory by default). Sort entries alphabetically
if none of -cftuvSUX nor --sort.
  Mandatory arguments to long options are mandatory for short options too.
 -a, --all
 do not ignore entries starting with .
 -A, --almost-all
 do not list implied . and ..
 --author
 with -l, print the author of each file
  -b, --escape
 print octal escapes for nongraphic characters
 --block-size=SIZE
 use SIZE-byte blocks.  See SIZE format below
```

Create Directory

Now let us create a directory named "Linux_tutorial". We will use this directory to complete the rest of the tutorial. The command that is used to create directory is 'mkdir'. Therefore, to create the directory "Linux_tutorial", we type:

```
$ mkdir Linux_tutorial
```

The above command creates a directory named "Linux_tutorial" in the current directory (e.g. our home directory in this case. Do you remember our home directory?).

Since the directory has been created, we would like to verify whether it is actually created or not. Any guesses how do we do that? Just using an 'ls' command as follows:

```
$ ls
```

```
$ ls
Linux_tutorial
```

Now you can see a folder named "Linux_tutorial" is listed on the screen. You can also check the contents of this newly created folder by using the following command:

```
$ls Linux_tutorial
```

Remember, Linux is case sensitive. "Linux_terminal" is different from "linux_terminal". Since "Linux_tutorial" is empty, nothing will be displayed. Go ahead and try the command.

Print Working Directory

Before we change our working directory to Linux_tutorial, let's check our present working directory first by typing 'pwd' command which stands for 'print working directory'.

```
$ pwd
```

```
$ pwd
/home/ngsguide
```

As can be observed from the output, 'pwd' prints the complete path [starting from root (/)] of current working directory or just the directory where the user is, at present.

From the above output in the box, it can be inferred that the directory name "**/home/ngsguide**" means "the directory named **ngsguide** is our current directory, which is in the directory named **home**, which in turn is in the directory named **root (/)**." All directories on a Linux file system are subdirectories of the root directory (http://www.computerhope.com/unix/ucd.htm).

Change Directory

In order to use the directory we had just created in the steps above, we need to navigate into that directory to make it our current working directory where we will complete the remaining steps of this tutorial. 'cd' (change directory) command is used to change your current directory. This 'cd' command can be used to traverse through the hierarchy of Linux file system (http://www.computerhope.com/unix/ucd.htm).

To change into Linux_tutorial directory and make it our working directory, we would use the command:

```
$ cd Linux_tutorial
```

Check which our current working directory is. Use 'pwd' command again and observe the difference between the output before and after changing the directory:

```
$ pwd
```

```
$ pwd
/home/ngsguide/Linux_tutorial
```

Download Data

Since our current directory is empty, we need some data files to go ahead with the rest of the tutorial. The 'wget' command[i] is used to download files from your Linux terminal provided you already have web link of that file. By using 'wget' we can also download FASTQ files from the public databases by providing the exact URL of the file. For instance, let's download a FASTQ file such as ERR000001_1.fastq.gz from EBI (http://www.ebi.ac.uk/) database.

[i] In general, 'wget' application should already exist on your Linux system. In case 'wget' is missing, it can be easily installed by using any of the following commands depending on your Linux distribution:

```
$ yum install wget
```

or

```
$ apt-get install wget
```

```
$ wget ftp://ftp.sra.ebi.ac.uk/vol1/fastq/
ERR000/ERR000001/ERR000001_1.fastq.gz
```

```
$ wget "ftp://ftp.sra.ebi.ac.uk/vol1/fastq/ERR000/ERR000001/ERR000001_1.fastq.gz"
--2016-02-18 08:32:45-- ftp://ftp.sra.ebi.ac.uk/vol1/fastq/ERR000/ERR000001/ERR000001_1.fastq.gz
       => "ERR000001_1.fastq.gz"
Resolving ftp.sra.ebi.ac.uk... 193.62.192.7
Connecting to ftp.sra.ebi.ac.uk|193.62.192.7|:21... connected.
Logging in as anonymous ... Logged in!
==> SYST ... done.    ==> PWD ... done.
==> TYPE I ... done.  ==> CWD (1) /vol1/fastq/ERR000/ERR000001 ... done.
==> SIZE ERR000001_1.fastq.gz ... 31131066
==> PASV ... done.    ==> RETR ERR000001_1.fastq.gz ... done.
Length: 31131066 (30M) (unauthoritative)
100%[==========================================================================>]
31,131,066 316K/s in 97s
2016-02-18 08:34:39 (314 KB/s) - "ERR000001_1.fastq.gz" saved [31131066]
```

Once you enter the above mentioned command, some download information is displayed on the screen and the FASTQ file is saved as ERR000001_1.fastq.gz in your current folder (Linux_tutorial). Alternately, you can also download this FASTQ file directly by pasting this link (ftp://ftp.sra.ebi.ac.uk/vol1/fastq/ERR000/ERR000001/ERR000001_1.fastq.gz) into your web browser such as Mozilla or Internet Explorer.

Confirm whether FASTQ file is downloaded or not by using:

```
$ ls -lh
```

```
$ ls -lh
total 30M
-rw-rw-r-- 1 ngsguide ngsguide 30M Feb 18 08:34 ERR000001_1.fastq.gz
```

Note: Here we have used an additional option for 'ls' which is '-h'. To check why we have used '-h' option type -

```
$ man ls
```

From the above 'ls -lh' command we can see that ERR000001_1.fastq.gz is successfully downloaded in a compressed format having .gz as an extension. The size of this compressed file is about 30Mb. Downloading FASTQ files in a compressed form is a healthy practice as this compression reduces the size of these files substantially therefore it will definitely save some download time for you when the file size is huge.

File Compression

'gzip' command is used to compress as well as uncompress/decompress all the files with .gz file extension.
To uncompress ERR000001_1.fastq.gz, type:

```
$ gzip -d ERR000001_1.fastq.gz
```

Here '-d' means uncompress or decompress. Now to check if the file is uncompressed, type:

```
$ ls -lh
```

```
$ ls -lh
total 130M
-rw-rw-r-- 1 ngsguide ngsguide 130M Feb 23 02:59 ERR000001_1.fastq
```

Notice that .gz extension from ERR000001_1.fastq.gz is gone and the file size is also increased to 130Mb as compared to 30Mb in the compressed FASTQ file.

Similarly, you can compress the ERR000001_1.fastq file using the 'gzip' command as follows:

```
$gzip ERR000001_1.fastq
```

Confirm whether the file is compressed again:

```
$ ls -lh
```

```
$ ls -lh
total 130M
-rw-rw-r-- 1 ngsguide ngsguide 30M Feb 23 02:59 ERR000001_1.fastq.gz
```

Observe that the compressed file with.gz file extension is created again. To proceed further with the tutorial, decompress this .gz file again with 'gzip' command as explained above.

Display the Contents of a File

The 'cat' command which stands for concatenate is used to display the contents of a file. To display the contents of ERR000001_1.fastq file, type:

```
$ cat ERR000001_1.fastq
```

This command will display contents of your FASTQ file on the terminal. However, in larger files such as FASTQ, most of the output will scroll up the screen with only the last part that can be accessed on the terminal. Type the above 'cat' command to get the output.

Other handy commands to quickly check the contents of bigger files are 'head' and 'tail' commands.

```
$ head ERR000001_1.fastq
```

This will display the first 10 (the default number) lines from the file.

```
$ head ERR000001_1.fastq
@ERR000001.1 IL2_62_3_1_346_881/1
GAACTAAGTGAACTGAAACATCTAAGTAACTTAAGG
+
IIIIIIIIIIIIIIIIIIIIIIIIIIIIIIIIIIII
@ERR000001.2 IL2_62_3_1_583_614/1
GATCCTACTATTACAATAATGCATTACAATATTACT
+
IIIIIIIIIIIIIIIIIIIIIIIIIIIIIIIIIIII
@ERR000001.3 IL2_62_3_1_389_877/1
GGGAGACAATGCAGAGGTTGAAAGATGTATCTGAAA
```

To print the first 15 lines, type:

```
$ head −15 ERR000001_1.fastq
```

```
$ head −15 ERR000001_1.fastq
@ERR000001.1 IL2_62_3_1_346_881/1
GAACTAAGTGAACTGAAACATCTAAGTAACTTAAGG
+
IIIIIIIIIIIIIIIIIIIIIIIIIIIIIIIIIIII
@ERR000001.2 IL2_62_3_1_583_614/1
GATCCTACTATTACAATAATGCATTACAATATTACT
+
IIIIIIIIIIIIIIIIIIIIIIIIIIIIIIIIIIII
@ERR000001.3 IL2_62_3_1_389_877/1
GGGAGACAATGCAGAGGTTGAAAGATGTATCTGAAA
+
IIIIIIIIIIIIIIIIIIII>IIICIIIII-IIIII
@ERR000001.4 IL2_62_3_1_284_606/1
TTAACGACCGTACCGAAAGTGGACTTAAGTAGTATG
+
```

Similarly, to print the last 10 and last 15 lines of a file, use:

```
$ tail ERR000001_1.fastq
$ tail -15 ERR000001_1.fastq
```

The output of the above examples shows the reads from the FASTQ file. For more details on FASTQ format, please refer to Chapter 3 of this book. Briefly, each read in a FASTQ file consists of 4 lines as shown below.

```
@ERR000001.1 IL2_62_3_1_346_881/1              #a unique sequence identifier
GAACTAAGTGAACTGAAACATCTAAGTAACTTAAGG           #the sequence
+                                              #a '+' that may be followed by the sequence
identifierIIIIIIIIIIIIIIIIIIIIIIIIIIIIIIIIIIII #the quality values
```

Count the Number of Lines

The 'wc'- word count command is used to count the number of lines in a file. To count the number of lines in a FASTQ file, type:

```
$ wc -l ERR000001_1.fastq
```

```
$ wc -l ERR000001_1.fastq
4683176 ERR000001_1.fastq
```

'-l' (L in lower case) option is used to print the newline counts. It can be observed that the FASTQ file ERR000001_1.fastq has 4683176 lines.

Search a Pattern

The 'grep' command is used to search patterns in an input file. When 'grep' finds a pattern match in a line, it prints the line to standard output. For example, to find a string of nucleotides "CCCCCTTAAAAA" in FASTQ file, type the following command:

```
$ grep "CCCCCTTAAAAA" ERR000001_1.fastq
```

```
$ grep "CCCCCTTAAAAA" ERR000001_1.fastq
AGTTTTTCATCAACCCCCTTAAAAAAATACATAGTT
CCCTTACCGGCCGTCCCCCTTAAAAAGAGGGCCGAC
TCATCAACCCCCTTAAAAAAATACATAGTTCTTAGG
AGTTTTTCATCAACCCCCTTAAAAAAATACATAGTT
```

All lines containing the nucleotide string "CCCCCTTAAAAA" are printed. However, this does not print the identifier for each sequence. We shall revisit this pattern searching again in the later examples.

Combine Multiple Commands Together

The pipes denoted by '|' are used to connect multiple commands together. By means of pipes, the standard output of one command is redirected as the standard input for another command. Count the number of reads from a FASTQ file as follows:

```
$ grep "@ERR000001" ERR000001_1.fastq | wc -l
```

```
$ grep "@ERR000001" ERR000001_1.fastq | wc -l
1170794
```

In the above example, the output of 'grep' command is fed to 'wc' command. We know that each read consists of a unique identifier that starts with a "@" symbol followed by an ID (e.g. ERR000001). The grep command extracts the lines with the pattern "@ERR000001" from a FASTQ file and this output is fed to 'wc' which counts the number of lines having the pattern "ERR000001". This implies that there are 1170794 reads in this FASTQ file. This can be easily verified by a simple formula:

$$\textit{No. of reads in a fastq file} = \frac{\textit{No. of reads in a fastq file}}{4}$$

where 4 = number of lines per read.

Converting a FASTQ File into a Tabular Format

Although raw data among NGS platforms are different, there are available tools to convert data to the *de facto* standard—FASTQ format. Occasionally it is very helpful to have the data in a tabular form (http://www.ark-genomics.org/events-online-training-eu-training-course/linux-and-bioinformatics).We can convert the FASTQ file into a tabular format by using 'cat' command and combining it with 'paste' command.

```
$  cat ERR000001_1.fastq | paste - - - - >
ERR000001_1_tab.txt
```

Here, the 'cat' command reads the FASTQ file and sends the output as an input for 'paste' command via pipe '|'. In the example above, each '-' reads a line from the standard input. Therefore, '- - - -' means read 4 lines and paste them next to each other (http://www.ark-genomics.org/events-online-training-eu-training-course/linux-and-bioinformatics; http://www.theunixschool.com/2012/07/10-examples-of-paste-command-usage-in.html). Make sure that there is a space between each '-'. The output of paste command is redirected to a new file named ERR000001_1_tab.txt. Verify if ERR000001_1_tab.txt is created:

```
$ ls
```

```
$ ls
ERR000001_1.fastq
ERR000001_1_tab.txt
```

By using 'ls', you can see that there are two files now. Now check the contents of the newly created file ERR000001_1_tab.txt. Just display first 10 lines:

```
$ head ERR000001_1_tab.txt
```

```
$ head ERR000001_1_tab.txt
@ERR000001.1 IL2_62_3_1_346_881/1 GAACTAAGTGAACTGAAACATCTAAGTAACTTAAGG + IIIIIIIIIIIIIIIIIIIIIIIIIIIIIIIIIIII
@ERR000001.2 IL2_62_3_1_583_614/1 GATCCTACTATTACAATAATGCATTACAATATTACT + IIIIIIIIIIIIIIIIIIIIIIIIIIIIIIIIIIII
@ERR000001.3 IL2_62_3_1_389_877/1 GGGAGACAATGCAGAGGTTGAAAGATGTATCTGAAA + IIIIIIIIIIIIIIIIIIII>IIICIIIII-IIIII
@ERR000001.4 IL2_62_3_1_284_606/1 TAACGACCGTACCGAAAGTGGACTTAAGTAGTATG + IIIIIIIIIIIIIIIIIIIIIIIIIIIIIIIIIII&
@ERR000001.5 IL2_62_3_1_480_810/1 GGTTTGCTTCAAGAATAGCTTTGGTTTGTAAAGGTT + IIIIIIIIIIIIIIIIIIIIIIIIIIIIIIIIIIII
@ERR000001.6 IL2_62_3_1_576_286/1 GATTTGTCAATCACTCGTGTTCCTTCCTATGTTTGT + IIIIIIIIIIIIIIIIIIIIIIIIIIIIIIIIIIII
@ERR000001.7 IL2_62_3_1_641_293/1 GGAAATGAAGGAAATGGAATTGCGTATTGTTGAATC + IIIIIIIIIIIIIIIIIIIIIIIIIIIIIII2IIII
@ERR000001.8 IL2_62_3_1_801_750/1 GGGATTTTAAAATTATTATTATATTTAAGAATAAGA + IIIIIIIIIIIIIIIIIIIIIIIIIIIIIIIIIIII
@ERR000001.9 IL2_62_3_1_386_889/1 TTATGTAGTACCTTTGTAATTATAATCATGATGATA + IIIIIIIIIIIIIIIIIIIIIIIIIIIIIIIIIIII
@ERR000001.10 IL2_62_3_1_866_369/1 GTCTTGAGTGAAGTTAAGGCCGAAGGCTTTGACAAA + IIIIIIIIIIIIIIIIIIIIIII<IIIIIII-IIEI
```

From the output it can be seen that the FASTQ file has now been converted into a tabular format. Now each line represents a single read (identifier, sequence, etc., all in one line) as compared to the earlier FASTQ format where 4 lines represented a single read.

Putting up all your data in a tabular format has its own advantages. For instance, look at the previous pattern searching example again. The command is same except that in the current example we have used the tabular formatted file.

```
$ grep "CCCCCTTAAAAA" ERR000001_1_tab.txt
```

```
$ grep "CCCCCTTAAAAA" ERR000001_1_tab.txt
@ERR000001.161625 IL2_62_3_27_924_80/1  AGTTTTTCATCAACCCCCTTAAAAAAATACATAGTT  +  IIIIIIIIIIIIIIIIIIIIIIIIIIIIIIIIAIII
@ERR000001.317933 IL2_62_3_54_744_131/1  CCCTTACCGGCCGTCCCCCTTAAAAAGAGGGCCGAC  +  IIIIIIIIIIII:IIII4I%.III63I5->2-**,
@ERR000001.570976 IL2_62_3_101_217_616/1 TCATCAACCCCCTTAAAAAAATACATAGTTCTTAGG  +  IIIIIIIIIIIIIIIIIIIIIIIIIIIIIIIIII3/
@ERR000001.751210 IL2_62_3_133_648_714/1 AGTTTTTCATCAACCCCCTTAAAAAAATACATAGTT  +  IIIIIIIIIIIIIIIIIIIIIIIIIIIIIIIIIIII
```

Now it is much easier to identify the reads having a string "CCCCCTTAAAAA" as well as their identifiers.

Pattern Matching Using Awk

Since we have data in a tabular format, it is more convenient to use 'awk' for enhanced data retrieval and text manipulation tasks. An awk script searches for lines in a file that comprises of given patterns (https://www.chemie.fu-berlin.de/chemnet/use/info/gawk/gawk_3.html). The syntax of a typical awk command is given below:

```
awk '/pattern to search/ {Actions}' filename
```

This means that awk will read each individual line in a file and if the line matches the pattern that is being searched, the action will be performed.

For example, we can also use awk to search the string "CCCCCT TAAAAA" as follows:

```
$ awk '/CCCCCTTAAAAA/ {print $0}' ERR000001_1_tab.txt
```

```
$ awk '/CCCCCTTAAAAA/ {print $0}' ERR000001_1_tab.txt
@ERR000001.161625 IL2_62_3_27_924_80/1  AGTTTTTCATCAACCCCCTTAAAAAAATACATAGTT + IIIIIIIIIIIIIIIIIIIIIIIIIIIIIIIIAIII
@ERR000001.317933 IL2_62_3_54_744_131/1  CCCTTACCGGCCGTCCCCCTTAAAAAGAGGGCCGAC + IIIIIIIIIIII:IIII4I%.III63I5->2-**,
@ERR000001.570976 IL2_62_3_101_217_616/1 TCATCAACCCCCTTAAAAAAATACATAGTTCTTAGG + IIIIIIIIIIIIIIIIIIIIIIIIIIIIIIIIII3/
@ERR000001.751210 IL2_62_3_133_648_714/1 AGTTTTTCATCAACCCCCTTAAAAAAATACATAGTT + IIIIIIIIIIIIIIIIIIIIIIIIIIIIIIIIIIII
```

In this example, /CCCCCTTAAAAA/ is a pattern (enclosed between forward slashes '/') whereas 'print $0' is the action used to print all the lines which match a given pattern. awk works by knowing the concepts of "file", "record" and "field". In an 'awk' data file each line represents one record, and only one record is operated by 'awk' at a time. Also, each record comprises of fields, that are separated by spaces or tabs (default separators of awk). As a result, the 1st field or column can be accessed with $1, 2nd field or column with $2, and so on. $0 means the full record or the entire file (http://www.arunviswanathan.com/content/ppts/awk_intro.ppt).Therefore, to only print the identifiers (1st column) and sequences (3rd column) for a given pattern, type:

```
$ awk '/CCCCCTTAAAAA/ {print $1 "\t" $3}'
ERR000001_1_tab.txt
```

```
$ awk '/CCCCCTTAAAAA/ {print $1 "\t" $3}' ERR000001_1_tab.txt
@ERR000001.161625     AGTTTTTCATCAACCCCCTTAAAAAAATACATAGTT
@ERR000001.317933     CCCTTACCGGCCGTCCCCCTTAAAAAGAGGGCCGAC
@ERR000001.570976     TCATCAACCCCCTTAAAAAAATACATAGTTCTTAGG
@ERR000001.751210     AGTTTTTCATCAACCCCCTTAAAAAAATACATAGTT
```

The above awk command reads all lines and prints only the 1st and 3rd columns from the file that contain "CCCCCTTAAAAA" pattern. Since the 3rd column consists of actual sequences in the tabular file (http://www.ark-genomics.org/events-online-training-eu-training-course/linux-and-bioinformatics), it may be convenient to use conditional pattern matching by using awk. For example, to find out which sequences have non-standard nucleotide bases such as "N", type:

```
$ awk '{if($3~"N") print $1 "\t" $3}'
ERR000001_1_tab.txt
```

```
$ awk '{if($3~"N") print $1 "\t" $3}' ERR000001_1_tab.txt
@ERR000001.562  AATGCTGAGGTANNTNANTTNNAGATACAACTAANT
@ERR000001.1144 TGTAACAAGTAANNCNANGTNNGTGCCATCTCTCNC
@ERR000001.1167 TTAAGTTGCTCCNNGNTNTTNNTAATGGCCTTCTNT
@ERR000001.1746 GGTATCACTTATNNCNCNTANNAGCCCAGCGGCGNT
@ERR000001.1754 TGACCCGGAAAANNANANTTNNATATTCTGCTGGNA
@ERR000001.1990 TCGTTAGTAAACNNCGANATNNTACGTGGCTGTTNT
@ERR000001.2014 TACGTGACGAACNNGNCNATNNCGTAGCCGATGANC
@ERR000001.2172 ATAAATTTGATCNNANGNAGNNCGAGGCGTTCCGNT
@ERR000001.2206 TAGAGAATGGTTNNCTGNAGNNCATAAAAGAGAGNT
@ERR000001.2414 TTTATACTTAGANNCATNTANNTTTAATCCCATCNT
```

In the above command, an "if" condition has been used. Simply, it means if any line in the 3rd column (sequence) has "N", then print its identifier (column 1) and sequence (column 3) only. Here "~" (tilde) is used for explicit pattern-matching expressions whereas "\t" means the separator between 1st and 3rd column should be a tab. (Note: Only few lines are displayed on the terminal.)

Sort and Extract Unique Sequences

We know that there are 1170794 sequences in the ERR000001_1_tab.txt file. There are chances that there might be duplicate sequences in the file even though they may have unique identifiers. To extract a set of unique sequences only, we first need to sort the file as follows:

```
$ sort -k3,3 ERR000001_1_tab.txt > ERR000001_
1_tab.srt
```

Now check the contents of ERR000001_1_tab.srt by printing the first 10 lines of this file:

```
$ head ERR000001_1_tab.srt
```

```
$ sort -k3,3 ERR000001_1_tab.txt > ERR000001_1_tab.srt

$ head ERR000001_1_tab.srt

@ERR000001.1000779 IL2_62_3_174_440_448/1    AAAAAAAAAAAAAAAAAAAAAAAAAAAAAAAAAAAA  +  II4II.
C1-:>75@+95/1;5+692+1+('<,$$.,

@ERR000001.1000782 IL2_62_3_174_606_366/1    AAAAAAAAAAAAAAAAAAAAAAAAAAAAAAAAAAAA  +
I<3I8I3E2I+*<<4)6.-0,<-(-2I,3/+*-3$7

@ERR000001.1000796 IL2_62_3_174_778_763/1    AAAAAAAAAAAAAAAAAAAAAAAAAAAAAAAAAAAA  +
FII?I;F51=7.>E>)I-1-51,>.-*9+,;'8+,'

@ERR000001.1000820 IL2_62_3_174_690_272/1    AAAAAAAAAAAAAAAAAAAAAAAAAAAAAAAAAAAA  +  :E41626
,A*>2&:5+/0.,+,&,$-/&6'&.$.&%

@ERR000001.1000826 IL2_62_3_174_544_885/1    AAAAAAAAAAAAAAAAAAAAAAAAAAAAAAAAAAAA  +
0H1@84:5878*()/*0-;*(%@(+')+')2)&+#2

@ERR000001.1000827 IL2_62_3_174_202_947/1    AAAAAAAAAAAAAAAAAAAAAAAAAAAAAAAAAAAA  +  5:/:/?:<
/5)89++)+22(0,&/4.+(1')%&++&

@ERR000001.1000844 IL2_62_3_174_510_789/1    AAAAAAAAAAAAAAAAAAAAAAAAAAAAAAAAAAAA  +  0EIIIIII
IIIIIII?II;I7H0I>,5I,2@';3(

@ERR000001.1000850 IL2_62_3_174_265_909/1    AAAAAAAAAAAAAAAAAAAAAAAAAAAAAAAAAAAA  +  .IIIIII
IIIIIIIIIIIIIIIIIIIIIIIIIII

@ERR000001.1000854 IL2_62_3_174_506_596/1    AAAAAAAAAAAAAAAAAAAAAAAAAAAAAAAAAAAA  +  6CICIFII
I685IFEHICH:=59146>I*8456273

@ERR000001.1000857 IL2_62_3_174_240_819/1    AAAAAAAAAAAAAAAAAAAAAAAAAAAAAAAAAAAA  + )
IIIHIIIII7>IAIHEIII4/-5CC5.?0<)400B
```

From the above output we can see that the 3rd column (i.e. the sequences) is sorted in the ascending order, showing many duplicate sequences. In the sort command, '-k' is used to specify the column on the basis of which the sorting will be carried out. The format of '-k' is: '-km,n' where 'm' is the starting column and 'n' is the end column (http://www.theunixschool.com/2012/08/linux-sort-command-examples.html). Since in our case, the sorting is based on the 3rd column only, we specify '-k3,3'.

Now to sort and extract only the unique sequences in a single step, we add one more parameter to the above command:

```
$ sort -k3,3 -u ERR000001_1_tab.txt >
ERR000001_1_tab.srt.unq
```

Check the number of lines in the newly created file ERR000001_1_tab.srt.unq:

```
$ wc -l ERR000001_1_tab.srt.unq
```

You can see that there are 1023945 lines now as compared to 1170794 lines in the original file ERR000001_1_tab.txt.

```
$ sort -k3,3 -u ERR000001_1_tab.txt > ERR000001_1_tab.srt.unq
$ wc -l ERR000001_1_tab.srt.unq
1023945 ERR000001_1_tab.srt.unq
```

Note: In the above command, we have used an additional parameter '-u' which means print only unique lines.

Convert Reads into FASTA Format Sequences

Most of the softwares or tools recognize sequences in FASTA format. Therefore, it will be very helpful to convert the reads into FASTA formatted sequences. The steps that will carry out this conversion are described below:

In the first step, extract the unique identifier (1st column) and the sequence (3rd column) from the tabular file created in the previous steps and save the output as a new file.

```
$ awk '{print $1 "\t" $3}' ERR000001_1_tab.txt
> ERR000001_1_allseqs.txt
```

Check if the new file ERR000001_1_allseqs.txt is created by using 'ls' command as described earlier in this chapter. Now verify if only two columns (identifier and sequence) are extracted from the tabular file ERR000001_1_tab.txt.

```
$ head ERR000001_1_allseqs.txt
```

```
$ head ERR000001_1_allseqs.txt
@ERR000001.1     GAACTAAGTGAACTGAAACATCTAAGTAACTTAAGG
@ERR000001.2     GATCCTACTATTACAATAATGCATTACAATATTACT
@ERR000001.3     GGGAGACAATGCAGAGGTTGAAAGATGTATCTGAAA
@ERR000001.4     TTAACGACCGTACCGAAAGTGGACTTAAGTAGTATG
@ERR000001.5     GGTTTGCTTCAAGAATAGCTTTGGTTTGTAAAGGTT
@ERR000001.6     GATTTGTCAATCACTCGTGTTCCTTCCTATGTTTGT
@ERR000001.7     GGAAATGAAGGAAATGGAATTGCGTATTGTTGAATC
@ERR000001.8     GGGATTTTAAAATTATTATTATATTTAAGAATAAGA
@ERR000001.9     TTATGTAGTACCTTTGTAATTATAATCATGATGATA
@ERR000001.10   GTCTTGAGTGAAGTTAAGGCCGAAGGCTTTGACAAA
```

Now add FASTA file identifier “>” at the beginning of each line by using ‘sed’ command:

```
$ sed -i ‘s/^/>/’ ERR000001_1_allseqs.txt
```

‘sed’ is a stream editor that is used to carry out simple text manipulation tasks on an input file (https://www.gnu.org/software/sed/manual/sed.html). In the above example, “s” represents the substitution action. The forward slashes (“/”) are delimiters. The “^” matches the null string at start of the pattern space, i.e. whatever appears next to the “^” must appear at the beginning of the pattern space (http://www.computerhope.com/unix/used.htm). “>” is the character that has to be added. Finally, “-i” option means to reflect the changes in the file ERR000001_1_allseqs.txt.

To summarize, we are invoking a ‘sed’ command which adds “>” at the beginning of each line (represented by “^”) in a file.

To verify that ">" has been added at the beginning of each line, use the 'head' command again:

```
$ head ERR000001_1_allseqs.txt
```

```
$ head ERR000001_1_allseqs.txt
>@ERR000001.1    GAACTAAGTGAACTGAAACATCTAAGTAACTTAAGG
>@ERR000001.2    GATCCTACTATTACAATAATGCATTACAATATTACT
>@ERR000001.3    GGGAGACAATGCAGAGGTTGAAAGATGTATCTGAAA
>@ERR000001.4    TTAACGACCGTACCGAAAGTGGACTTAAGTAGTATG
>@ERR000001.5    GGTTTGCTTCAAGAATAGCTTTGGTTTGTAAAGGTT
>@ERR000001.6    GATTTGTCAATCACTCGTGTTCCTTCCTATGTTTGT
>@ERR000001.7    GGAAATGAAGGAAATGGAATTGCGTATTGTTGAATC
>@ERR000001.8    GGGATTTTAAAATTATTATTATATTTAAGAATAAGA
>@ERR000001.9    TTATGTAGTACCTTTGTAATTATAATCATGATGATA
>@ERR000001.10  GTCTTGAGTGAAGTTAAGGCCGAAGGCTTTGACAAA
```

You can now clearly see that ">" has been added before each line. Still this is not a proper FASTA formatted file.

```
$ awk'{ print $1, "\n" $2}' ERR000001_1_
allseqs.txt > ERR000001_1_allseqs.fasta
```

Here we use awk again to convert it into a FASTA file. In this example, we are simply printing the 1st column (the identifier), followed by a newline character (\n) and finally the sequence itself (2nd column). "\n" will allow the sequence to be printed on a new line and the output is redirected to a new file (ERR000001_1_allseqs.fasta). To check whether we have the FASTA file or not, type:

```
$ head ERR000001_1_allseqs.fasta
```

```
$ head ERR000001_1_allseqs.fasta
>@ERR000001.1
GAACTAAGTGAACTGAAACATCTAAGTAACTTAAGG
>@ERR000001.2
GATCCTACTATTACAATAATGCATTACAATATTACT
>@ERR000001.3
GGGAGACAATGCAGAGGTTGAAAGATGTATCTGAAA
>@ERR000001.4
TTAACGACCGTACCGAAAGTGGACTTAAGTAGTATG
>@ERR000001.5
GGTTTGCTTCAAGAATAGCTTTGGTTTGTAAAGGTT
```

Now, you can see the fasta formatted file is created.

Write a Shell Script to Split Sequences into Individual Files

Until now all sequence reads are in just one file and sometimes there might be a requirement to separate these sequences into individual files. Now we make use of shell scripting to split each sequence into individual files. In a typical UNIX-like system (including Linux), Shell has been instrumental in bridging between the user and the computer. Shell is a command interpreter that interprets user instructions to Kernel for further execution. There are many types of Shell in Linux such as: Bourne Shell (SH), C Shell (CSH), Korn Shell (KSH), TC Shell (TCSH) and Bourne Again Shell (BASH). The latter one (BASH) is the most popular Shell because it incorporates useful features from the KSH and CSH. A Shell is not only an excellent command line interpreter, but also has scripting features that allows automation of tasks that would otherwise require lot of steps. You can visit http://linuxcommand.org/lc3_lts0010.php for a more detailed explanation about Shell. To give instruction to the Shell, we shall use a text input and output environment called Terminal (http://linuxcommand.org/lc3_lts0010.php).

To start writing a shell script, we need to use a text editor. There are a few text editors that we can use such as 'pine', 'pico' and 'vi'. In this chapter the vi (pronounced as: vee ay) text editor will be used. It is a screen-oriented text editor originally created for the Unix operating system. The vi editor is the most common text editor that Linux users use to edit text files or scripts. To start using the vi editor, simply type 'vi' followed by the text file name that you want to edit or a new text file that you want to create. vi editor has two modes, namely the "command mode" (the default mode when the file is opened or created) and the insert 'i' mode (you need to be in insert mode to write the shell script).

Now let's start creating a shell script to separate a multiFASTA file into individual FASTA files. Type:

```
$ vi split.sh
```

This will create and open a new file named split.sh (assuming that no file with the same name exists already). However, as mentioned previously, this file will be opened in a command mode. To change it into insert mode, press 'i' or 'a' key to activate the insert mode and type the following to create your first shell script:

```
#!/bin/bash
INPUT_FILE=$1
PREFIX=$2
csplit -z $INPUT_FILE '/^>/' '{*}' --suffix="%02d.fasta" --prefix=$PREFIX -s
```

The first line is the statement to tell the Shell to use BASH as the default shell to run the scripts.

Like the other command line program, BASH allows the user to pass some values to the script from the command line. This value is called an argument. The argument is stored in variable with a number in the order of the argument starting at 1 (e.g. $1, $2 , $3, etc.). The second and third lines are the statement to "hold" the first and second arguments that will be passed by the user in variables (i.e. INPUT_FILE and PREFIX).

The last statement is where we split the multisequenceFASTA file into individual FASTA files by using "csplit" program. The csplit takes seven arguments:

1. -z : remove empty output files
2. $INPUT_FILE : the multisequencesFASTA file
3. '/^>/' : the regular expression statement to find line that starts with '>' character
4. '{*}' : repeat the previous pattern as many times as possible
5. --suffix :add suffix to each individual file (%20d will be replaced by a sequence number)
6. --prefix : add prefix to each individual file
7. --s : do not print counts of output file sizes

Once you are done with typing the script in vi, you need to close and save the file before executing the script. Follow these two steps to exit from vi and save the file:

1. Press the 'Escape' key to quit from the insert mode.
2. Type ':wq' and press 'Enter' key to save (w) and quit (q) from vi text editor.

Refer to http://ryanstutorials.net/linuxtutorial/vi.php to get acquainted with vi.

Changing File Permissions

Now that the script is ready, we have to change it into an executable format by changing its permissions before we can actually execute it.

Linux has inherited from UNIX the concept of ownerships and permissions for files. This is basically because it was conceived as a networked system where different people would be using a variety of programs, files, etc. Obviously, there's a need to keep things organized and secure. We don't want an ordinary user using a program that could potentially trash the whole system. There are security and privacy issues here as well. Below are some examples of file permission attributes on Linux:

rwxrwxrwx: Three sets of rwx. The leftmost set pertains to the owner, the middle set is for the group, and the rightmost set is for others; rwx stands for read (r), write (w), execute (x); the dash (-) means no permission.

Other examples are:

rwx------: Only the owner can read, write, and execute.
rw-r--r--: Everyone can read, and the owner can also write.
rw-------: Only the owner can read and write.
r--r--r--: Everyone can read.

Permissions can also be expressed numerically, where read (r) is equal to 4, write (w) is equal to 2, execute (x) is equal to 1, and no permission is equal to 0. Therefore, rwxrwxrwx is equal to 777, rwx------ is equal to 700, rw-r--r-- is equal to 744, rw------- is equal to 600, and r--r--r-- is equal to 444.

To change the permission and file/folder ownership, the following commands can be used:

- chmod: To change file/folder permission, example: chmod 755 split.sh or chmod +x split.sh
- chown: To change file ownership, example: chown user 2 split.sh
- chgrp: To change folder ownership, example: chgrp group2 split.sh

Run the Bash Script

Before we test the script, it is good to test with a sample from the multiFASTA file first before you proceed with the actual file. Recall that your multiFASTA file (ERR000001_1_allseqs.fasta) has 1170794 sequences. Now let's take some sample sequences from this file to test the bash script by using the following command:

```
$ head -n 24 ERR000001_1_allseqs.fasta > sample.fasta
```

The command above will create a sample multiFASTA file with 12 sequences. Now, run the shell script:

```
$ chmod +x split.sh
$ ./split.sh sample.fasta ERR000001_1_
$ ls
ERR000001_1_00.fasta
ERR000001_1_01.fasta
ERR000001_1_02.fasta
ERR000001_1_03.fasta
ERR000001_1_04.fasta
ERR000001_1_05.fasta
ERR000001_1_06.fasta
ERR000001_1_07.fasta
ERR000001_1_08.fasta
ERR000001_1_09.fasta
ERR000001_1_10.fasta
ERR000001_1_11.fasta
```

The first command is used to give executable permission to the shell script file (i.e. split.sh), while the second command is used to run the shell script. As a gentle reminder, the shell script requires two arguments: the multiFASTA file (e.g. sample. fasta file consisting of all the FASTA sequences) and output file prefix (e.g. ERR000001_1_). After the shell script file is executed, it generates several FASTA files with one sequence per file.

Summary

This tutorial provides a brief introduction to some of the widely used Linux commands that will help a potential user to quickly generate some statistics about their data. There is a lot of help available online for these and many more Linux commands. A large number of free tutorials on the usage of these commands are also easily accessible online. Having such a basic skill in Linux is very important in order to efficiently organize, manipulate and analyze any kind of biological data generated by high throughput technologies. Although it may sound demanding initially, the effort is rewarding as you may have seen in this chapter. Most of the bioinformatics tools are developed to work on the Linux system and furthermore the majority of High Performance Computing systems are using Linux as the operating system. Finally, for more advanced analysis of bioinformatics data, a user may want to consider learning at least one programming language (e.g. PERL or Python).

Chapter 3

Inspection of Sequence Quality

Kwong Qi Bin, Ong Ai Ling, Heng Huey Ying and
Martti T. Tammi

Biotechnology & Breeding Department, Sime Darby Plantation
R&D Centre, Selangor, 43400, Malaysia.

Glossary of Terms

FASTQ: Text-based nucleotide sequence with its quality score. Line 1 is the FASTA identifier, line 2 is the nucleotide sequence, line 3 starts with a '+' followed by the optional FASTA identifier, and line 4 represents the quality score.

FASTA: Text-based nucleotide sequence without its quality score, which is just line 1 and 2 of FASTQ. It has a header that starts with ">" and the next line is the sequence.

Kmer: Nucleotide sub-sequence that is made up of a fixed number of K bases.

PCR: Polymerase chain reaction is a molecular biology method that is used to amplify a DNA fragment to multiple copies.

GC: Guanine-cytosine content within a sequence.

Introduction

The adage 'garbage in, garbage out' serves as an important reminder to users of NGS technologies to be careful about the quality of sequence data that are used for analyses. Although NGS is a powerful technology that allows us to acquire important biological information of a species such as its genome, its accuracy depends on the raw sequenced data. Similar to any high throughput system, NGS is bound to have some errors during sequencing. To assess sequence quality, Phred score was introduced and it is basically a probability measurement of a wrong base call. It is calculated as below:

$$Q = -10\log_{10}(e),$$ where *e* is the estimated wrong base call probability

Therefore, Q scores of 10, 20 and 30 represent 1 incorrect base call in 10, 100 and 1000 bases, respectively. This score, however, is encoded differently in American Standard Code for Information Interchange (ASCII) code in different Illumina systems. The earlier Illumina (1.3, 1.5) systems use ASCII of 64–126 to represent the Q score of 0–62, which is actually the Phred score +64. In the more recent Illumina (1.8 and 1.9) systems, ASCII of 33–93 (Phred + 33) is used. Therefore, when working with raw **FASTQ** reads, one needs to know the right sequence quality encoding method prior to plotting a distribution of the quality of sequenced data. Low quality bases and in some cases, entire reads can then be removed according to criteria set by the researcher. Besides quality of individual bases, it is also important to remove adaptor sequences and any suspected contaminant sequences. The pre-processing of raw sequence reads to ensure only high quality data are used for subsequent steps such as alignment or assembly is the focus of the practical in this chapter.

FastQC

FastQC[1] is a quality control tool for NGS data. It is useful in summarizing the quality of sequencing and detecting potential problems. The program can be downloaded at http://www.bioinformatics.babraham.ac.uk/projects/fastqc/.

Installation Step in Linux Environment

The software has included a wrapper script called 'fastqc' which is the easiest way to start the program. The wrapper is in the top level of the FastQC installation. In order to make this file executable:

```
$ wget http://www.bioinformatics.babraham.ac.uk/projects/fastqc/fastqc_v0.11.9.zip
$ unzip fastqc_v0.11.9.zip
# Note that a folder named FastQC is generated.
$ cd FastQC/
# It is useful to look inside the file INSTALL.txt to see some useful instructions on how to use the software.
$ chmod 755 fastqc
```

Once you have done that you can run it directly by typing:

```
$ ./fastqc
```

An error might show if you cannot view the graphical interface of FastQC as shown in Figure 1. To fix this error, the user needs to ensure they have X11 display.

Figure 1. Main graphical interface of FastQC.

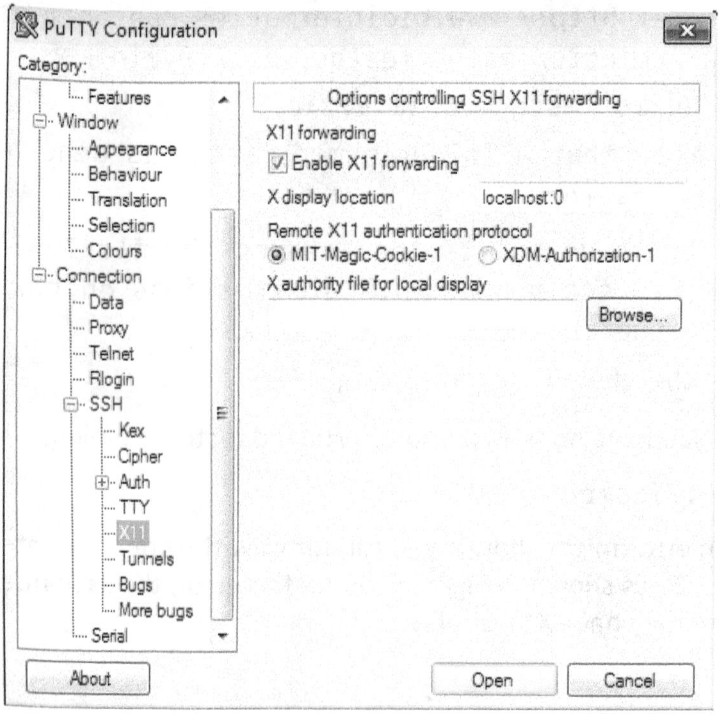

Figure 2. Screenshot for Putty's configuration to enable Xming X for any visualization purposes.

If you are running Putty in Windows, Xming X server is required to view the output figure files from the server side. This program can be found at http://sourceforge.net/projects/xming/. After that, you will need to run the Putty program with the X11 option enabled, as shown in Figure 2.

In Mac, if you are connecting to a server, the command is usually

```
$ ssh –X <user>@<server IP>
```

In some cases, X11 might need to be downloaded and installed. For Mac, you can download and install from this link http://apple-x11.en.softonic.com/mac/download. From our experience, Linux usually comes with X11 being installed. If there are still problems, kindly refer to the system administrator of your organization for help.

You may also place a link in /usr/local/bin to be able to run the program from any location:

```
$ sudo ln -s /path/to/FastQC/fastqc /usr/
local/bin/fastqc
```

Download Datasets

The datasets can be downloaded at http://bioinfo.perdanauniversity.edu.my/infohub/display/NPB/Index

Three sample files have been provided, namely "good_seq.fastq", which is a FASTQ file with good quality, "bad_seq.fastq", a file with bad quality and "contaminated.fq", a file with many different contaminants.

We will be running this step in the FastQC folder. You can also try to create a separate folder and run this program in the folder.

```
$ mkdir test1
$ cd test1
```

Before running any analysis, let us find out the number of sequences in the FASTQ file. Taking "good_seq.fastq" as an example:

```
$ wc -l good_seq.fastq
```

This command tells us that there are 1 million lines in the files, and since FASTQ has 4 lines for a single entry, hence there are ...0,000 sequences in the FASTQ file.

...command below will show us the first record.

```
... 4 good_seq.fastq
...:192:#0/1
...ATCAGACCAAAATTCTTAAGT

...;9;5?<=?@??><=<;8
```

...VQB1:1:10:72:192:#0/1 and it rep... which is unique for each sequence.

The second line is the bases, third line is mainly just a separator, and the fourth line is the quality scores. To run FastQC for all the example files, use the following command:

```
$ fastqc good_seq.fastq bad_seq.fastq
contaminated.fq
```

Three output files will be generated for each of the input files:

```
good_seq_fastqc.zip, bad_seq_fastqc.zip and
contaminated_fastqc.zip.
```

If these files are not automatically unzipped, you can manually unzip them using the following command:

```
$ unzip '*.zip'
contaminated_fastqc.zip
```

Let us have a look at the one of the unzipped contents:

```
$ cd good_seq_fastqc
$ ls –lah
```

To view the output (Figure 3), we will be using Firefox to view the report in HTML format. Besides Firefox, Chrome, Safari and other browsers can also be used. For most operating systems, installation for Firefox can be found at https://support.mozilla.org/en-US/products/firefox/download-and-install.

```
$ firefox fastqc_report.html
```

Another option is to transfer the output files using WinSCP from the server to your personal computer (Windows-based) before viewing it. WinSCP can be downloaded from http://winscp.net/eng/index.ph

```
total 464K
drwxrwxr-x 4 student student 512  May 24 15:01 .
drwxrwxr-x 3 student student 8.0K May 24 15:30 ..
-rw-rw-r-- 1 student student 7.0K May 24 15:02 fastqc_data
-rw-rw-r-- 1 student student 3.6K May 24 15:02 fastqc.fo
-rw-rw-r-- 1 student student 216K May 24 15:02 fastqc_re
drwxrwxr-x 2 student student 512  May 24 15:01 Icons
drwxrwxr-x 2 student student 8.0K May 24 15:01 Images
-rw-rw-r-- 1 student student 518  May 24 15:02 summar
```

Figure 3. List of output from the program

In the unzipped folder of the output files, there are also two text files, which are 'summary.txt' and 'fastqc_data.txt'. These files contain the raw statistics that are used to generate the HTML reports. Take a look at it if you are interested. For the purpose of this tutorial, we will only focus on the more user-friendly HTML report.

Overall, it is easy to locate potential problems in the FASTQ files by looking at the summary column in the HTML file. The summary has 11 categories that show various aspects relevant for sequence quality inspection (Table 1).

Table 1. Various analysis modules incorporated in the FastQC program.

Analysis Modules	**Definitions**
Basic statistics	General statistics and some background information regarding the input file
Per base sequence quality	Bases' quality values across all the reads of the input FASTQ file
Per tile sequence quality	The average quality scores from each tile across all the bases
Per base sequence content	Percentage of A, C, G, T across the FASTQ reads
Per base **GC** content	GC content across the FASTQ reads, for each base position
Per sequence GC content	Average GC distribution over all sequences, and provided a comparison of it with a normal distribution
Per base N content	Percentage of N base calls at each position across the FASTQ reads
Sequence length distribution	Summary on length distribution for the FASTQ reads, useful after trimming reads
Sequence duplication levels	Summary of the counts for every sequence in the FASTQ file, useful in detecting biased enrichment problems such as **PCR** over amplification
Overrepresented sequences	Frequency summary of sequences, useful in detecting and classifying contaminants in sequencing, for example PCR primers
Adapter content	Cumulative plot of the fraction of reads where the sequence library adapter sequence is identified

Besides each of the categories, there are symbols that represent their results:

 Represents problem in this category

 Represents acceptable in this category

 Represents warning in this category

The FastQC program is capable of detecting problems regarding base and sequence quality, base content, **Kmer** frequency, GC content, sequence length and duplication and contaminant/adapter. Undeniably, all of these problems will affect the quality of assembly of mapping, but from our experience, the two main problems are the base quality and adapter. It is also worth noting that the other problems might need to be solved at the library preparation and sequencing stage, and is out of the scope of this tutorial. This tutorial will focus primarily on base quality and adapter problem.

From our previous example run, we would have generated an example result with good sequence quality result in the good_sequence_short_fastqc folder. This is the most important figure generated from the program. Figure 4 shows how a good sequence quality FastQC result should look like.

The y-axis of the figure represents the Q score, and the x-axis is the position of the base in a raw read. On top of the figure, we know that the encoding type is Illumina 1.9. Usually, the quality of the bases deteriorates towards the end of the read, with the forward read showing better quality than the reverse read. Overall, the quality of this dataset is defined as good, because the box plots which represent the base quality were all in the green region (score >28). All the bases, on average, has base quality of >30. A warning will be issued if the median for any base with score <25, and a failure if <20. We will look at other figures next.

Figure 5 shows the quality score distribution over all sequences. The average quality per read is actually very high, at 32. A good dataset will have a single peak located around score 30. Warning is given when the mean quality is <27, failure at <20.

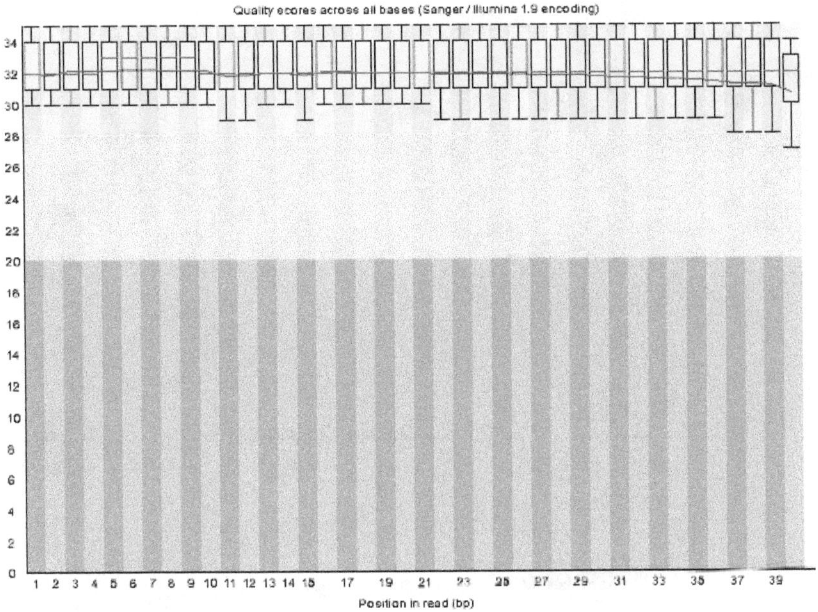

Figure 4. Example result for good sequence quality scores.

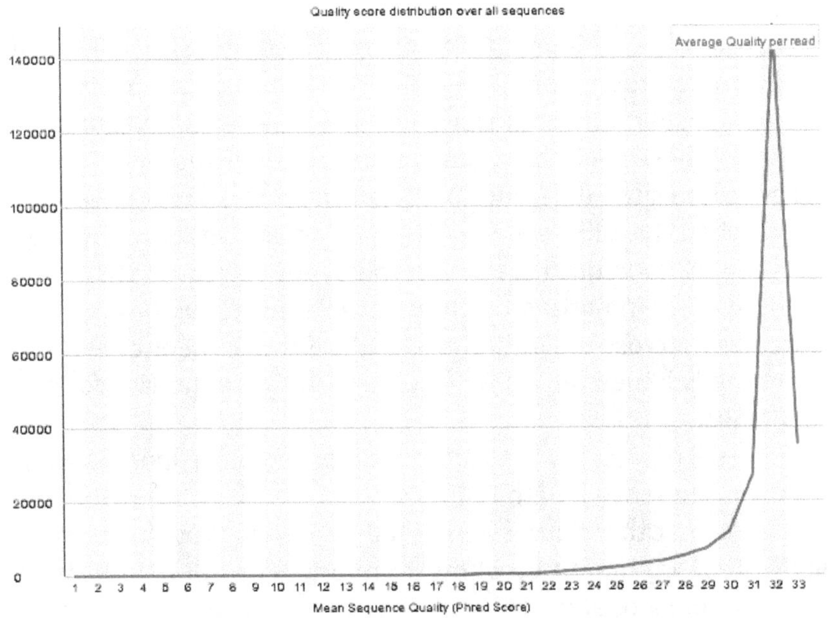

Figure 5. Example results for average quality score for all sequences.

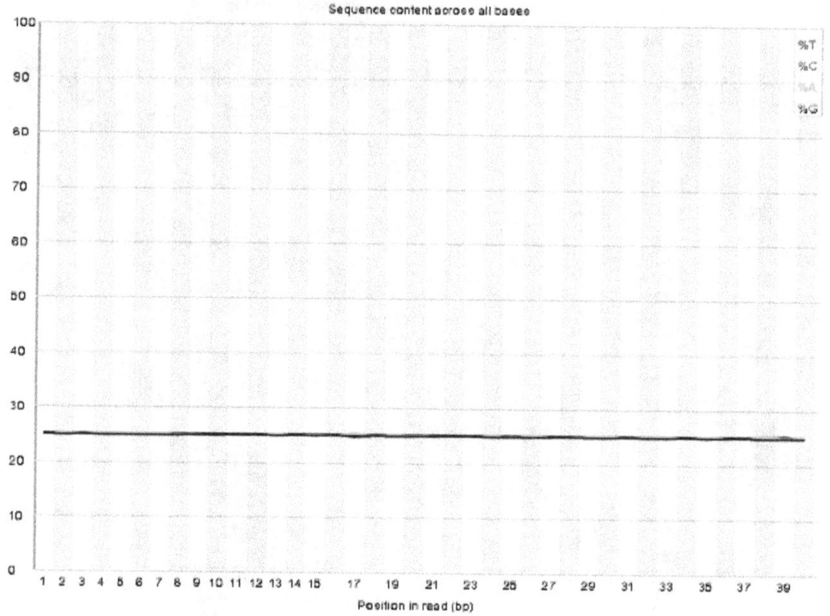

Figure 6. Example result for bases content of all sequences.

Figure 6 represents the sequence content across all bases. Ideally, there should not be any base preference at any positions. In reality, however, this is almost never the case. At the beginning of the reads, there might be some base bias and fluctuations. Warning will be issued if the difference between any of the bases to be >10%, failure when this difference reaches 20% at any position. Even for a dataset with good quality, this test might not necessary pass.

Figure 7 summarizes the number of ambiguous bases, represented as N across the entire raw reads. If the number of N is >5%, a warning is issued, at >20%, failure. In this case, no N is found in the dataset.

Figure 8 represents the distribution of sequence lengths. In this case, the sequence length is 40 bp.

There are other figures generated by FastQC, but above are the figures that we find to be of most importance with regards to the base/sequence quality problem.

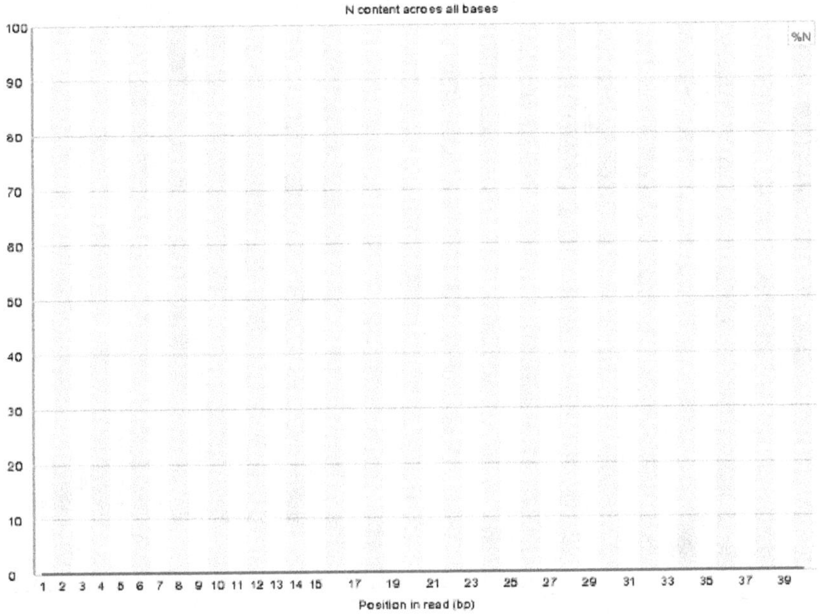

Figure 7. Example result for N (ambiguous) content across all sequences.

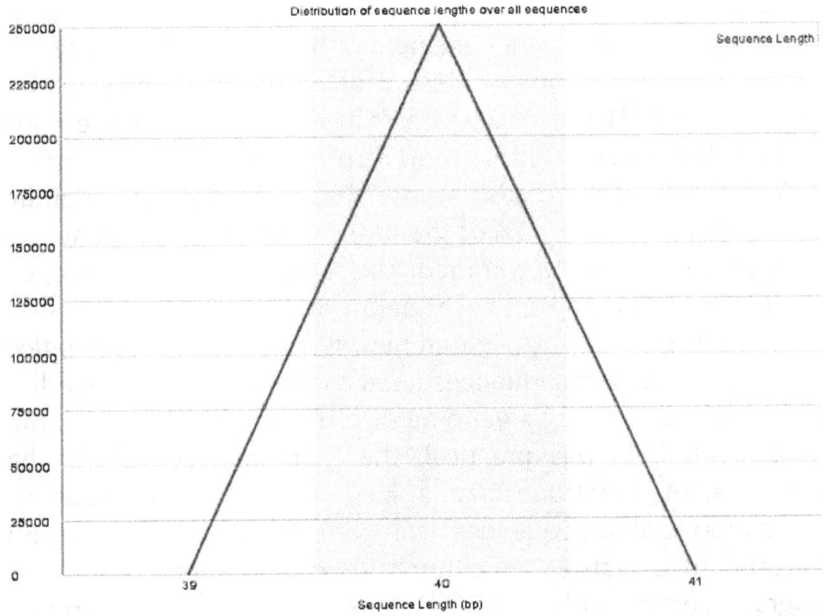

Figure 8. Example result for distribution of sequence length over all sequences.

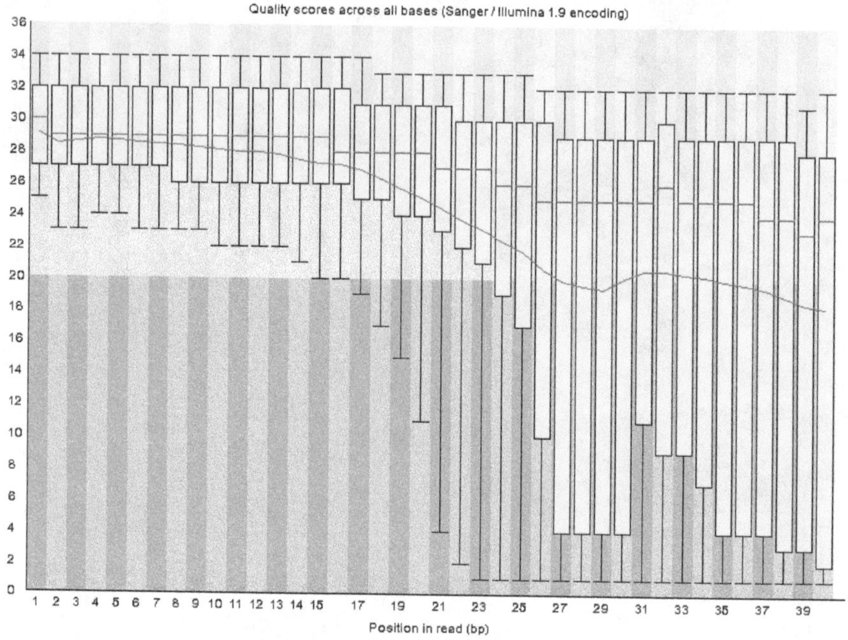

Figure 9. Example result for bad sequence quality scores.

Let us take a look at an example of bad sequence quality. The files are in the bad_sequence_fastqc folder. The “Per base sequence quality” figure (Figure 9) best describes the quality problem and will therefore be selected for more explanation.

The quality of the reads dropped drastically after base number 15. Approximately 20–40% of the bases were in the red zone, and these bases need to be trimmed. This is not an absolute rule, but it is fairly common to trim bases below quality of 20 or 25.

Another problem highlighted here is sequence contamination. These contaminated sequences need to be dealt with properly in the case of assembling a genome or else a lot of false genes may be generated. In this practical, the “contaminated.fq” file has adapter contamination. Take a look at the result under the “Over represented sequences” category. Most of the time, over-represented sequences are either primers or adapters. The FastQC program comes with a folder (/path/to/FastQC/Configuration/contaminant_list.txt) that lists down most of the primers, adapters

Sequence	Count	Percentage	Possible Source
GATCGGAAGAGCACACGTCTGAACTCCAGTCACATCACGA	6276	6.276	TruSeq Adapter, Index 1 (100% over 40bp)
AATGATACGGCGACCACCGAGATCTACACTCTTTCCCTAC	6274	6.274	Illumina Single End PCR Primer 1 (100% over 40bp)
CAAGCAGAAGACGGCATACGAGATCGTGATGTGACTGGAG	6252	6.252000000000001	Illumina PCR Primer Index 1 (100% over 40bp)
CAAGCAGAAGACGGCATACGAGATACATCGGTGACTGGAG	6192	6.192	Illumina PCR Primer Index 2 (100% over 40bp)
GATCGGAAGAGCGGTTCAGCAGGAATGCCGAGACCGATCT	6142	6.142	Illumina Paired End PCR Primer 2 (100% over 40bp)

Figure 10. Example result for contamination problem.

or other contaminants. Regardless, it is usually good to contact your sequencing service provider to find out the primers and adapters used, in case the list provided by FastQC is not up to date.

The result for contamination problem is in the contaminated_fastqc folder. For this case, we will be focusing on the "Overrepresented sequences" table (Figure 10). As you can see from under the "Possible Source" column, the problem is caused by Paired End PCR primers and adapters.

FastQC is a comprehensive program that perform quality inspection of sequencing data. For more information regarding FastQC can be found at http://www.bioinformatics.babraham.ac.uk

Fastx-toolkit & FASTQ Processing Utilities

Upon quality check of your dataset, the next step would be to get the data ready for further mapping work. Fastx-toolkit[2] is a collection tools for FASTQ files preprocessing. It can be downloaded from http://hannonlab.cshl.edu/fastx_toolkit/download.html.

Installation Step in Linux Environment

```
$ wget http://hannonlab.cshl.edu/fastx_tool-
kit/fastx_toolkit_0.0.13_binaries_Linux_2.6_
amd64.tar.bz2
```

```
$ tar -xjf fastx_toolkit_0.0.13_binaries_
Linux_2.6_amd64.tar.bz2
```

A bin directory will be created and the programs will be stored in it. It can be copied to /usr/bin/ directory. Alternatively, just list the full path when running the program. $ sudo cp /path/to/bin/* /usr/bin/

In previous steps, we have highlighted some problems regarding sequencing quality. After detecting these problems, we can now trim or remove bases or sequences that are of low quality. For this purpose, we will look at three tools from the Fastx toolkit. The first tool that we will look at is the FASTQ Quality Filter, which filters sequences based on their quality. For this purpose, we will be using the file "bad_seq.fastq". Please note that if the Illumina encoding is >= 1.8, one needs to provide a –Q33 option in using this command. For Illumina encoding of <= 1.5, this option is not required.

```
$ fastq_quality_filter -i bad_seq.fastq -q 25
-p 80 -o bad_seq.fastq.filtered –Q33
```

Flag:

-Q33: Illumina 1.9 encoded (i.e. ASCII code = Phred + 33)
-i: Input file name
-q: minimal quality of base to keep
-p: minimal percentage of bases that must have at least q quality
-o: output file name

Next, we will look at FASTQ Quality Trimmer, which shortens reads in a FASTQ file based on the quality.

```
$ fastq_quality_trimmer -i bad_seq.fastq.
filtered -t 25 -o bad_seq.fastq.trimmed –Q33
```

Flag:

-t: Quality threshold - nucleotides with lower than the threshold quality will be trimmed (from the end of the sequence). Take note that the –q argument in fastq_quality_filter is different from the –t argument here. In –q argument, the entire sequence will be removed for those that do no pass the threshold whereas the –t argument will only trim bases from the end of the sequence.

After correction, let us run FastQC again.

```
$ fastqc bad_seq.fastq.trimmed
```

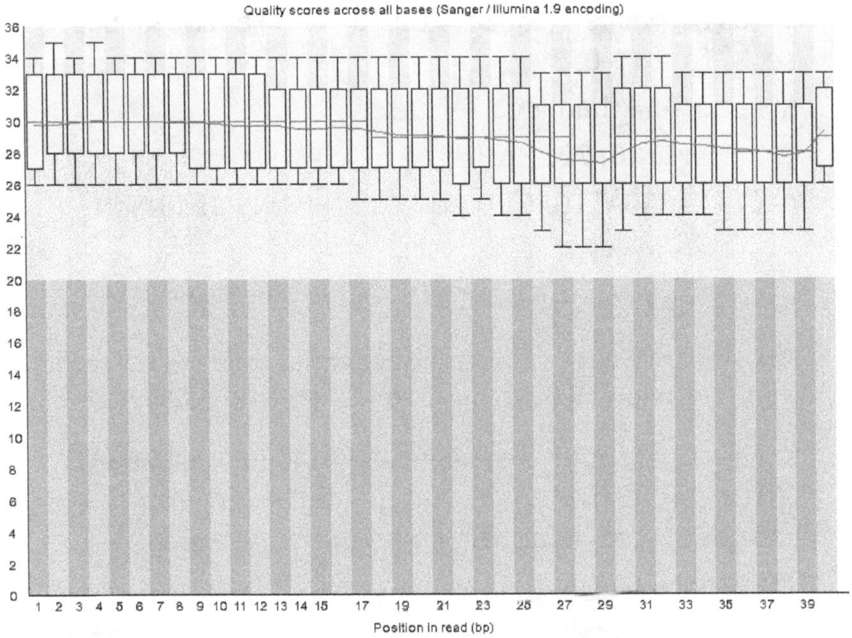

Figure 11. FastQC result after quality trimming.

As we can see in Figure 11, the quality of the input file has improved drastically. Although some of the bases still fall into the yellow zone, most of the low quality ones have been removed.

Another option is to remove the entire low-quality short read by setting a more stringent cutoff for the quality scores. However, we strongly do not recommend doing this.

```
$ fastq_quality_filter -q 20 -p 80 -i bad_seq.
fastq -o bad_sequence.txt.filtered -Q33
```

This is because some assembly programs are "picky", they do not allow for raw reads of different length or different number of reads in the paired-end FASTQ files. `Fastq_quality_filter` is likely to remove some paired end reads, thus turning them into singletons. For cases like this, the solution is to substitute the low quality bases with Ns. This can be done using `fastq_masker`.

```
$ fastq_masker -q 20 -i bad_seq.fastq -o bad_seqN.fastq -Q33
$ head -n4 bad_seqN.fastq
@HWQB1:1:10:72:192:#0/1
AACTTCTGGGATTGAGTTCNNNNNNNNNNNNNNNNNNNNN
+
A>>A@D>69=7=<9<;<:<"""#""%"""$""""%""%%"
```

Now, to remove the sequencing adapters' problem that we have encountered before, we can make use of the FASTQ Clipper tool.

```
$ fastx_clipper -a CAAGCAGAAGACGGCATACGAGAT
CGTGATGTGACTGGAG -i contaminated.fq -o
contaminated_adapter_remove.fq -v -Q33
```

Or

```
$ fastx_clipper -a CAAGCAGAAGA -i contaminated.
fq -o contaminated_adapter_remove.fq -v -Q33
```

Flag:

-a: adapter sequence (referred to Figure 10 as example)
-v: for verbose mode

It is rather cumbersome to remove all the adapters this way, as we need to remove the adapters one by one, therefore we can use another tool set, known as FASTQ processing utilities[3] by Erik Aronesty. See below for the commands to install and run:

```
$wget --no-check-certificate https://storage.googleapis.com/google-code-archive-downloads/v2/code.google.com/ea-utils/ea-utils.1.1.2-537.tar.gz
$ tar -xzvf ea-utils.1.1.2-537.tar.gz
$ cd ea-utils.1.1.2-537
$ sudo make install
```

The adapter file has been given as "adapters_give.txt". This file is in the FASTA format.

```
$ ea-utils.1.1.2-537/fastq-mcf adapters_give.txt
contaminated.fq -o contaminated_adapter_remove.fq
```

Run FastQC again,

```
$ fastqc contaminated_adapter_remove.fq
```

From the result generated, we observed that most of the adapters have been removed. Take note on the changes in 'Sequence Length Distribution' as well.

In some cases, after correcting the raw reads, one might just want to convert FASTQ file to a FASTA file. This can be done using the fastq_to_fasta tool.

```
$ head -n8 bad_seq.fastq
@HWQB1:1:10:72:192:#0/1
AACTTCTGGGATTGAGTTCTCTGACCAGCCTGGTGCCTCG
+
A>>A@D>69=7=<9<;<:<"""#""%"""$"""""%""%%"
@HWQB2:1:10:72:192:#0/1
CGAGGGGGGTTTCAGGATACAGAGTTACTCAAACATACC
+
>9<A<:?=9C>8B>A<DBB;@""%#"""""""$"#"#$#""
$ fastq_to_fasta -i bad_seq.fastq -o bad_seq.fa
-Q33
$ head -n4 bad_seq.fa

>HWQB1:1:10:72:192:#0/1
AACTTCTGGGATTGAGTTCTCTGACCAGCCTGGTGCCTCG

>HWQB2:1:10:72:192:#0/1
CGAGGGGGGTTTCAGGATACAGAGTTACTCAAACATACC
```

Table 2. Content of fastx-toolkit.

Tools	Function
FASTQ-to-FASTA converter	Convert FASTQ files to FASTA files
FASTQ/A barcode splitter	Split files containing multiple samples
FASTQ collapser	Collapse identical sequences
FASTQ renamer	Rename the sequence identifiers
FASTQ/A reverse-complement	Generate reverse-complement of each sequence
FASTQ information	Chart quality statistics and nucleotide distribution
FASTA formatter	Change the width of sequences line in a FASTA file
FASTA nucleotide changer	Convert FASTA sequences from/to RNA/DNA

There are other useful tools in fastx-toolkit as well, which are listed in Table 2.

Conclusion

It is important to do sequence quality inspection to ensure a good and clean data for downstream analyses. Options available are to either omit the entire sequence, low quality bases or to treat low quality bases as Ns.

References

1. Andrews, S. *FastQC: a quality control tool for high throughput sequence data*, <http://www.bioinformatics.babraham.ac.uk/projects/fastqc> (2010).
2. Gordon, A. & Hannon, G. J. Fastx-toolkit. (2010).
3. Aronesty, E. *ea-utils*: Command-line tools for processing biological sequencing data, <https://github.com/ExpressionAnalysis/ea-utils> (2011).

Chapter 4

Alignment of Sequenced Reads

Akzam Saidin

Novocraft Technologies Sdn Bhd, Selangor, Malaysia.

Introduction

Next(second) generation sequencer platforms by Illumina, SoLiD and Ion Torrent generate a high throughput volume of short reads or paired reads. The reads generated from short reads sequencer platforms are typically >= 300 base pairs with low read error profile. Some of the latest sequencing technologies (third generation sequencers) based on single molecule system by Pacific Biosciences and Oxford Nanopore produce longer reads, in the range of 1,000–40,000 base pairs with a higher read error profile. Both generations of sequencers have their pros and cons and the choice of platform depends on the problem.

In a re-sequencing study, reads generated by the sequencing machine will need to be aligned to a reference sequence. This step is called reads alignment or mapping. By performing reads alignment to a reference sequence, researchers can perform genetic variants detection in their sequenced samples.

To perform read alignments to a reference sequence, an aligner software is used. An aligner software will read the sequence reads (i.e. in FASTQ format) and compare it to the reference sequence by using a mapping algorithm. The aligner will try to find highly similar sequence location in the reference. A simplified representation of read alignments is as shown below.

```
Reference     GGATCCATGCGTCCCAGGTCACGGGATCCATG CGTCCCAGGTCACG
                                              *

Read A              ATGCGTCCCAGGTCACGGGATCCATGCGTCC
Read B          ATCCATGCGTCCCAGGTCACGGGGTCCATGC
Read D                 CGTCCCAGGTCACGGGATCCATGCGTCCCAG
Read F                      CAGGTCACGGGATCCATGCGTCCCAGGTCAC
```

The first row represents the reference sequence and below are the aligned reads. Read A, D and F are a perfect match to the reference. Read B has single base difference to the reference, it contains a nucleotide base G instead of nucleotide base A in reference (location indicated by * symbol).

The reads alignment information are collected by an aligner and it is usually reported in an alignment file in Sequence Alignment/Map (SAM) format or in its binary format, Binary Alignment/Map (BAM).

There are a multitude of read aligners available, a comprehensive list can be found at EBI HTS Mapper page (http://www.ebi.ac.uk/~nf/hts_mappers/). In the following section, we will go through the basic alignment process for both short and long reads using BWA and novoAlign.

Practical

Short Reads Alignment

In this section we will perform reads alignment using BWA aligner.

Dataset

	Info	File(s)
Reference sequence	Escherichia coli K12 MG1655 (ENSEMBL)	ecoliK12MG1655_ensembl.fna

Read set	Illumina GAII (Run ID: ERR008613) Paired end reads subsampled to 10X coverage	GA2_R1.fastq, GA2_R2.fastq

The datasets can be downloaded at http://bioinfo.perdanauniversity.edu.my/infohub/display/NPB/Index

Software Requirements

Software	Version	URL
BWA[1]	0.7.13-r1126	https://github.com/lh3/bwa/
*Novoalign[2]	V3.04.04	http://www.novocraft.com/support/download/
*Novosort[3]	V1.03.09	http://www.novocraft.com/support/download/
SAMTOOLS[4]	1.3	http://www.htslib.org/download/
IGV[5]	2.3.60	https://www.broadinstitute.org/igv/

* alternative software(s)

Installation instructions for each software can be found on the download site.

Alignment Process

Optional: index reference sequence

It is preferred to index the reference fasta file, especially when you have multiple sequences as references. FAI index enables efficient access in the alignment file to arbitrary regions within those reference sequences.

Create reference sequence index

```
samtools faidx ecoliK12MG1655_ensembl.fna
```

this command will produce the following reference index file

```
ecoliK12MG1655_ensembl.fna.fai
```

Create BWA reference index

```
bwa index ecoliK12MG1655_ensembl.fna
```

this command will produce the following bwa index files

```
ecoliK12MG1655_ensembl.fna.amb
ecoliK12MG1655_ensembl.fna.ann
ecoliK12MG1655_ensembl.fna.bwt
ecoliK12MG1655_ensembl.fna.pac
ecoliK12MG1655_ensembl.fna.sa
```

Align reads to reference

Align reads with BWA

```
bwa mem ecoliK12MG1655_ensembl.fna GA2_R1.
fastq GA2_R2.fastq > aln-pe.bwa.sam 2> bwa.log
```

View SAM file

To view the SAM file on terminal

```
View BWA SAM file

less -S aln-pe.bwa.sam

@SQ  SN:GCA_000005845.2:Chromosome:1:
4641652:1  LN:4641652

@PG  ID:bwa  PN:bwa  VN:0.7.13-r1126 CL:bwa mem

ecoliK12MG1655_ensembl.fna GA2_R1.fastq GA2_
R2.fastq

EAS20_8_6_100_1000_1413  83 GCA_000005845.2:
Chromosome:1:4641652:1 3849483 60  100M =
3849364 -219

CGGCAGCGCCAGACAGAATGGCGTAAAGCGCGACAGT
TCGTCCGGCAATCCCAACTGGAGCCAGAGACTGATA
ACAAACAGCAGCAAGTACCAGACCAGA

F@>CCFFE/HFHHHHFFF5F@DFBED@CDBFCEHHDHDHH@
BF?BFHE5HHGHHG; ===6HHHHHHGHHHHEHHHHHHH
GIIHHHHGGHFFFFBFFFFBB  NM:i:0  MD:Z:100
AS:i:100  XS:i:0
```

```
EAS20_8_6_100_1000_1413  163 GCA_000005845.2:Ch
romosome:1:4641652:1 3849364 60  100M =
3849483 219

GAGAGCAATAAATCCACCGGATGATCGCGCCAGGTTTG
ACTGGCGATCAGCGCGATGGCGTTCATCAACGTCG
CAATCAGCGCCCCTTGCCAACCATAGT

AEGE>FHFHCEGG@EFHEHHHFEFCHHGF@HFHIDHGDHHHDH=
F?EEFH@BHHG>>F;F=FDCDFE6BBBFEEC7D6=D6E?:?GFEHBGGC
CGE@AFB NM:i:0 MD:Z:100 AS:i:100 XS:i:0
```

The header section is indicated by the @ symbol.

Tag	Description
HD	Header
• VN	SAM format version
• SO	Sorting order of alignments
SQ	Reference sequence dictionary (information)
• SN	Reference sequence name
• LN	Reference sequence length
PG	Program
• ID	Program record identifier
• PN	Program name
• VN	Program version
• CL	Command line

The alignment section consists of multiple TAB-delimited lines with each line describing an alignment. A better view of the fields is as shown in the transposed example below:

Field	Value
QNAME	EAS20_8_6_100_1000_1413
FLAG	83

RNAME	GCA_000005845.2:Chromosome:1:4641652:1
POS	3849483
MAPQ	60
CIGAR	100M
MRNM/ RNEXT	=
MPOS/ PNEXT	3849364
ISIZE/TLEN	-219
SEQ	CGGCAGCGCCAGACAGAATGGCGTAAAGCGCGACAGTTC GTCC GGCAATCCCAACTGGAGCCAGAGACTGATAAC AAACAGCAGCAAGTACCAGACCAGA
QUAL	F@>CCFFE/HFHHHHFFF5F@DFBED@ CDBFCEHHDHDHH@BF?BFHE5HHGHHG;===6HHHHHHG HHHHEHHHHHHHGIIHHHHGGHFFFFBFFFFBB
TAG(s)	NM:i:0 MD:Z:100 AS:i:100 XS:i:0

A basic explanation on the SAM format fields as seen in above

Field	Brief description
QNAME	Query template NAME
FLAG	bitwise FLAG
RNAME	Reference sequence NAME
POS	1-based leftmost mapping POSition
MAPQ	Mapping Quality
CIGAR	CIGAR string
RNEXT	Ref. name of the mate/next read
PNEXT	Position of the mate/next read
TLEN	Observed Template Length
SEQ	Segment SEQuence

QUAL	ASCII of Phred-scaled base QUALity+33
TAGs	Additional information tagged to alignment

A more detailed explanation can be found in the SAM Format specification document (http://samtools.github.io/hts-specs/).

The following fields are usually checked on to assess the reads alignment:

1. FLAG
   This field contains the flag number for types of reads alignment. For example, a read pair that is flagged as '2' is paired-end reads that are mapped properly. The flags can be checked using the picard explain flags tool at http://broadinstitute.github.io/picard/explain-flags.html
2. CIGAR
   The CIGAR string is a simplified sequence mapping representation. The string shows alignment from the aligner on the number of bases that aligns (match/mismatch) with the reference, deleted from the reference, insertions that are not in the reference and soft/hard clipping of the sequence reads from being aligned to the reference. A simple CIGAR table is provided, see Table 1. A complete CIGAR table can be found in SAM format documentation at http://samtools.github.io/hts-specs/).

Table 1. Simple CIGAR table.

Op	Description
M	alignment match (can be a sequence match or mismatch)
I	insertion to the reference
D	deletion from the reference
N	skipped region from the reference
S	soft clipping (clipped sequences present in SEQ)
H	hard clipping (clipped sequences NOT present in SEQ)

For example:

```
                    1          2        3        4
           1234567890123456 7890123456789012345678901234567890123456
Reference  GGATCCATGCGTCCCA GGTCACGGGATCCATGCGTCCCAGGTCACG
Read B       ATCCAT CGTCCCATGGTCACGGGGTCCATGC
```

Which will report:

```
POS      3
CIGAR    5M1D7M1I17M
```

The POS indicates the base position on the reference; in this example Read B starts at position 3 with 5 matches. At position 9, it has 1 deletion (highlighted yellow; not present in read sequence). 7 matches from position 10 before an insertion (highlighted green; not present in reference). Then it is followed by 17 matches inclusive of the mismatch bases in position 26.

3. QUAL
   QUAL is a value for how accurate each base in the query sequence (SEQ) is.
   Quality is calculated based on the probability that a base is wrong, p, using the Phred Quality score (http://en.wikipedia.org/wiki/Phred_quality_score) formula:

$$quality = -10\ log_{10}p$$

   In SAM format, the 'p' value is added with 33 (this is to enable the value to be within readable ASCII printing range). The QUAL field for SAM uses the following formula:

$$QUAL = (-10\ log_{10}p) + 33$$

4. MAPQ
   MAPQ is the quality value for mapping, rounded to the nearest integer. It uses the following formula:

$$MAPQ = -10\ log_{10}p\ Pr\{mapping\ position\ is\ wrong\}$$

It ranges from value 0 to 255. Do be careful and refer to the aligner documentation on how MAPQ should be interpreted because different aligners use different MAPQ values.

SAM to BAM conversion

Convert SAM to BAM

To convert SAM format to BAM format, we use SAMTOOLS.

```
samtools view -uS -o aln-pe.bwa.bam aln-pe.
bwa.sam
```

Sort BAM alignments

Sorting BAM file in general is a process to sort aligned reads based on the aligned position to the reference genome. A sorted BAM is usually a requirement for some, if not most analysis tools. Sorting reads to reference position helps in increasing efficiency in reading, processing and compacting the file size.

Sorting BAM file can be done using SAMTOOLS

1. **Sort alignment with reference coordinate order**

```
samtools sort -T aln.tmp.sort -o aln-pe.bwa_
sorted.bam aln-pe.bwa.bam
```

2. **Index alignment**

```
samtools index aln-pe.bwa_sorted.bam
```

This will produce BAM index file

```
aln-pe.bwa_sorted.bam.bai
```

3. **Mark/remove duplicates**

```
samtools rmdup aln-pe.bwa_sorted.bam aln-pe.
bwa_rmdup.bam 2> samtools_rmdup.log
```

Perform indexing again on the output BAM file

```
samtools index aln-pe.bwa_rmdup.bam
```

Alternative: novoAlign & novoSort

Create novoAlign Index

Create novoAlign reference index

```
novoindex ecoliK12MG1655_ensembl.idx
ecoliK12MG1655_ensembl.fna
```

This command will produce the novoindex file.

Align reads with novoAlign

```
novoalign -d ecoliK12MG1655_ensembl.idx -o
SAM -f GA2_R1.fastq GA2_R2.fastq > aln-pe.
novoalign.sam 2> novoalign.log
```

Convert SAM to BAM

```
samtools view -uS -o aln-pe.novoalign.bam aln-
pe.novoalign.sam
```

Piping tips: align reads and convert SAM to BAM in one command line

```
novoalign -d ecoliK12MG1655_ensembl.idx -o SAM
-f GA2_R1.fastq GA2_R2.fastq 2> novoalign.log |
samtools view -uS -o aln-pe.novoalign.bam -
```

Sort and remove duplicate reads using novoSort

```
novosort -i --md -o aln-pe.novosort.bam aln-
pe.novoalign.bam 2> novosort.log
```

Parameter	Description
-i	create output bam index
--md	mark duplicates

View BAM alignment with IGV

Run IGV

```
./igv.sh
```

On the top panel, click on

```
Genomes
Load genome(s) from file
```

```
go to the folder ecoliK12MG1655_ensembl.fna is
located and choose ecoliK12MG1655_ensembl.fna
```

then click on

```
File
Load from file
```

```
choose aln-pe.bwa_rmdup.bam and aln-pe.
novosort.bam
```

this will display the alignments for both bam files. A snapshot of it is provided in Figure 1. Zoom in to make the reads bases visible.

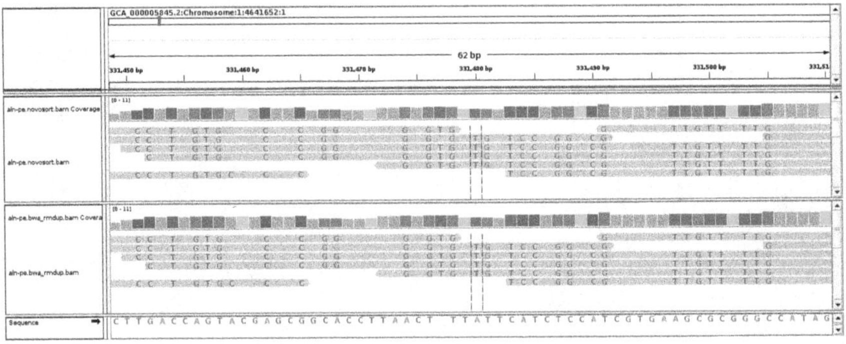

Figure 1. A zoomed in view of aligned reads in IGV.

References

1. Li, H. & Durbin, R. Fast and accurate short read alignment with burrows — Wheeler transform. *Bioinformatics* **25**, 1754–1760 (2009).

2. Hercus, C. Novoalign <www.novocraft.com/support/download/>Novocraft Technologies Sdn Bhd.
3. Hercus, C. Novosort <www.novocraft.com/support/download/>Novocraft Technologies Sdn Bhd.
4. Li, H. *et al.* The sequence alignment/map format and sAMtools. *Bioinformatics* **25**, 2078–2079 (2009).
5. Robinson, J. T. *et al.* Integrative genomics viewer. *Nature Biotechnology* **29**, 24–26 (2011).

Chapter 5

Establish a Research Workflow

Joel Low Zi-Bin and Heng Huey Ying

Biotechnology & Breeding Department, Sime Darby Plantation R & D Centre, Selangor, 43400, Malaysia.

Introduction

Planning is an important part of a project and this includes the creation of bioinformatics workflows. Before starting, it is imperative to be clear on the objective of the workflow. Planning a research workflow can be daunting for large experiments unless we break it down to smaller tasks, each with its own objectives. For each objective, there will be input and the corresponding output. When an output of a task is an input for another, one is able to connect the inputs and outputs of tasks to generate a workflow.

As an example, let us consider a research project that aims to study the genetic diversity of the bacteria *Escherichia coli*. To achieve this objective, a reference of the *E.coli* genome is required, which in turn requires the sequencing of *E.coli* DNA and its assembly. One major task that sits between DNA sequencing and genome assembly is the filtering of reads. To achieve this task, we will need a way to detect and remove poor quality reads. The program Trimmomatic[1] can do this. To determine if the removal was effective, we will want to compare the the quality of reads before and after quality trimming. We will need to employ a program like FastQC[2] to achieve this task. Therefore, one workflow that can be created is to combine filtering of reads and evaluate the quality of output.

While it is possible to run each step on the terminal, it would be far more efficient to automate the process in a workflow. The workflow can run each step without intervention from the user. A well written workflow will have the benefits of being reusable, documented and able to make multiple instance runs (i.e. having two or more workflows running at the same time). In the following practical, we will show you two ways to build a program workflow to filter reads using:

(1) Shell scripts
(2) Galaxy

Shell scripts are discussed in detail in Chapter 2. For the purpose of this chapter, a shell script is simply a compiled document of commands you would use as if writing directly on the command line.

Galaxy[3] is an open source, web-based platform for computational biomedical research. Galaxy's graphical user interface makes bioinformatics tools easily accessible for users with or without programming experience. The ability to create and save workflows in Galaxy enables users to repeat, share and learn complete computational analysis workflows. Users can use Galaxy via the public Galaxy's web server (http://usegalaxy.org), or set up their own Galaxy locally by downloading the Galaxy application (http://getgalaxy.org/).

Materials

- Illumina PE (short):
  - Source: https://www.ebi.ac.uk/ena/browser/view/ERX002508?show=reads
  - Info:
    - files are *gz* compressed.
    - quality value format: Sanger
    - # reads: 28,428,648
    - Read length (bp): 2×100

  - Insert size (bp): 215.4 ± 10.6
  - Illumina Genome Analyzer IIx
- Two sample files are available after unzipping the downloaded file:
  – s_6_1.fastq
  – s_6_2.fastq

Alternatively, you may download the files via command:

```
$ wget http://ftp.sra.ebi.ac.uk/vol1/run/
   ERR008/ERR008613/200x100x100-
   081224_EAS20_0008_FC30TBBAAXX-6.tar.gz
```

Unzip the downloaded file:

```
$  tar-xvf 200x100x100-081224_EAS20_0008_
   FC30TBBAAXX-6.tar.gz
```

- Software that will go into the workflow:

(a) FastQC — Detailed use and installation is covered in Chapter 3.
(b) Trimmomatic — For installation, download the binary version from the website and unzip the file. You will need java installed to run the trimmomatic-latest version 0.39.jar file. Detailed use is covered in Chapter 6. In this example, we used version 0.39.

```
$ wget http://www.usadellab.org/cms/uploads/
supplementary/Trimmomatic/Trimmomatic-0.39.zip
```

```
$ unzip Trimmomatic-0.39.zip
```

- Illumina adapters -> TruSeq2-PE.fa (provided by Trimmomatic in the "adapters" directory)
- For this practical, we will sample a smaller portion of the short reads for a quicker analysis run:

```
$ head -8000000 s_6_1.fastq |gzip -f >s_6_1.2M.
fastq.gz
```

```
$ head -8000000 s_6_2.fastq |gzip -f >s_6_2.2M.
fastq.gz
```

Shell Scripts

(1) You will first need to know the commands necessary for each step:

a. Run FastQC on each fastq file on the command line:

```
$ fastqc s_6_1.2M.fastq.gz s_6_2.2M.fastq.gz
```

Note: FastQC can accept gunzip compressed files directly.

b. Run Trimmomatic for both fastq files:

```
$ java -jar /path/to/Trimmomatic-.039/
trimmomatic-0.39.jar PE s_6_1.2M.fastq.gz
s_6_2.2M.fastq.gz s_6_1_paired.2M.fastq.gz
s_6_1_unpaired.2M.fastq.gz s_6_2_paired.2M.
fastq.gz s_6_2_unpaired.2M.fastq.gz
ILLUMINACLIP:/path/to/Trimmomatic-.039/
adapters/TruSeq2-PE.fa:2:30:10 LEADING:3
TRAILING:3 SLIDINGWINDOW:4:30 MINLEN:30
```

c. Run FastQC on each fastq output:

```
$ fastqc s_6_1_paired.2M.fastq.gz s_6_1_
unpaired.2M.fastq.gz s_6_2_paired.2M.fastq.
gz s_6_2_unpaired.2M.fastq.gz
```

Understanding the FastQC output is covered in Chapter 3, with some visual examples shown later in this chapter.

(2) Now, create a new file called **pipeline1.sh** at the command terminal by opening in **vi**:

```
$ vi pipeline1.sh
```

(3) Type in the shell interpreter location in the first line.

```
#!/bin/sh
```

(4) Write each command code per line into the shell script. The contents should look like this:

```
#!/bin/sh

# system time and date at start:
date

# start from the very beginning with the download of the files:
#wget http://ftp.sra.ebi.ac.uk/vol1/run/ERR008/ERR008613/200x100x100-081224_
EAS20_0008_FC30TBBAAXX-6.tar.gz

#Unzip the downloaded file

#tar -xvf 200x100x100-081224_EAS20_0008_FC30TBBAAXX-6.tar.gz

# sub-sampling to save time:
#head -8000000 s_6_1.fastq |gzip -f >s_6_1.2M.fastq.gz
#head -8000000 s_6_2.fastq |gzip -f >s_6_2.2M.fastq.gz

#Run FastQC on each fastq file on the command line:
fastqc s_6_1.2M.fastq.gz s_6_2.2M.fastq.gz

# Run Trimmomatic for both fastq files:
java -jar /path/to/Trimmomatic-.039/trimmomatic-0.39.jar PE s_6_1.2M.fastq.gz
s_6_2.2M.fastq.gzs_6_1_paired.2M.fastq.gz s_6_1_unpaired.2M.fastq.gz
s_6_2_paired.2M.fastq.gz s_6_2_unpaired.2M.fastq.gz
ILLUMINACLIP:/path/to/Trimmomatic-0.39/adapters/TruSeq2-PE.fa:2:30:10 LEADING:3
TRAILING:3 SLIDINGWINDOW:4:30 MINLEN:30

# Run FastQC on each fastq output:
fastqc s_6_1_paired.2M.fastq.gz s_6_1_unpaired.2M.fastq.gz s_6_2_paired.2M.fastq.gz
s_6_2_unpaired.2M.fastq.gz

# system time and date at end:
date
```

Notice that lines beginning with "#" are not commands. These are comment lines, used to document your script to make it easier to understand should you forget. Comment lines are ignored by the interpreter.

It is sometimes of interest to time the run of the script. In the script example above, the "date" function will take a snapshot of the date and time, which are strategically placed in the beginning and end of the script. Another way to time your script, is shown in Step 8.

(5) Save the file using the following **vi** command:

```
:wq
```

(6) Make sure the file is executable, by running the following command:

```
$ chmod 755 pipeline1.sh
```

(7) Run the program:

```
$ ./pipeline1.sh
```

(8) Alternatively, you can time the run of your script by adding the command "time" at the beginning of the script:

```
$ time ./pipeline1.sh
```

(9) Once the run is completed, the following output would be generated:

(a) Trimmomatic summary:

Read Pairs: 2000000 Both Surviving: 1498537 (74.93%) Forward Only Surviving: 218518 (10.93%) Reverse Only Surviving:
163048 (8.15%) Dropped: 119897 (5.99%).
The following files are generated:

(1) s_6_1_paired.2M.fastq.gz
(2) s_6_1_unpaired.2M.fastq.gz
(3) s_6_2_paired.2M.fastq.gz
(4) s_6_2_unpaired.2M.fastq.gz

(b) FastQC:

The summary results can be found in the summary.txt files, with details found in fastqc_data.txt. Graphical reports can be read by loading fastqc_report.html in a web browser.
The following files are generated:

(1) s_6_1.2M_fastqc/summary.txt
(2) s_6_2.2M_fastqc/summary.txt
(3) s_6_1.2M_fastqc/fastqc_report.html
(4) s_6_2.2M_fastqc/fastqc_report.html
(5) s_6_1_paired.2M_fastqc/summary.txt
(6) s_6_1_unpaired.2M_fastqc/summary.txt
(7) s_6_2_paired.2M_fastqc/summary.txt
(8) s_6_2_unpaired.2M_fastqc/summary.txt
(9) s_6_1.2M_fastqc/fastqc_report.html
(10) s_6_2.2M_fastqc/fastqc_report.html

(c) Total program workflow runtime: 6 minutes (may vary depending on the computer)

(10) If you had run the script above, and the following errors occur:

a. "...Permission denied"—Check if you had set the shell script to be executable. To fix it, do the following:

```
$ chmod 755 pipeline1.sh
```

b. "Unable to access..." or "...No such file or directory" — check if any of the files are in the same working directory. If you're accessing it from a different directory, do check if the file paths are correct.

c. "...command not found"— You may have forgotten to add "./" before the shell script command.

Note that a program command and its arguments are written in a single line. You can check this by running "less -N" to indicate line numbers when viewing the file:

```
$ less –N pipeline1.sh
```

You may decide to have the command broken up into different lines, which you can do by using the backslash, "\", at the end of each line. This could make a command with many arguments more readable. An example:

```
# Run Trimmomatic for both fastq files:
java -jar /path/to/Trimmomatic-.039/trimmomatic-0.39.jar PE \
s_6_1.2M.fastq.gz s_6_2.2M.fastq.gz \
s_6_1_paired.2M.fastq.gz \
s_6_1_unpaired.2M.fastq.gz \
s_6_2_paired.2M.fastq.gz \
s_6_2_unpaired.2M.fastq.gz \
ILLUMINACLIP:/path/to/Trimmomatic-.039/adapters/TruSeq2-PE.fa:2:30:10 \
LEADING:3 \
TRAILING:3
SLIDINGWINDOW:4:30 \
MINLEN:30
```

However, ensure that there isn't any whitespace after the back-slash, otherwise it will not work.

(11) The need for repeating the workflow may arise, and a few edits in the shell script is enough to get you going again. For example, you may want to try another parameter run, and still keep the previous command you used for documentation purposes, you can do the following in the script:

```
# Run Trimmomatic for both fastq files:
# first run: default
# java -jar /path/to/Trimmomatic-.039/trimmomatic-0.39.jar PE s_6_1.2M.fastq.gz
s_6_2.2M.fastq.gz
s_6_1_paired.2M.fastq.gz s_6_1_unpaired.2M.fastq.gz
s_6_2_paired.2M.fastq.gz s_6_2_unpaired.2M.fastq.gz
ILLUMINACLIP:/path/to/Trimmomatic-.039/adapters/TruSeq2-PE.fa:2:30:10 LEADING:3
TRAILING:3 SLIDINGWINDOW:4:30 MINLEN:30
# second run: no trim edges, qual >= 20:
java -jar /path/to/Trimmomatic-.039/trimmomatic-0.39.jar PE s_6_1.2M.fastq.gz
s_6_2.2M.fastq.gz
s_6_1_paired.2M.fastq.gz s_6_1_unpaired.2M.fastq.gz
s_6_2_paired.2M.fastq.gz s_6_2_unpaired.2M.fastq.gz
ILLUMINACLIP:/path/to/Trimmomatic-.039/adapters/TruSeq2-PE.fa:2:20:10
```

(12) Not all programs would alert users of its completion. For very long workflows, it is possible to add in simple alerts between steps to help in identifying steps with errors. This would save time rerunning steps that do work. One could add the following between the FastQC and Trimmomatic steps:

```
#Run FastQC on each fastq file on the command line:
fastqc s_6_1.2M.fastq.gz s_6_2.2M.fastq.gz
echo "FastQC on subsampled data complete. Running Trimmomatic now..."
# Run Trimmomatic for both fastq files:
java -jar /path/to/Trimmomatic-.039/trimmomatic-0.39.jar PE s_6_1.2M.fastq.gz
s_6_2.2M.fastq.gz
s_6_1_paired.2M.fastq.gz s_6_1_unpaired.2M.fastq.gz
s_6_2_paired.2M.fastq.gz s_6_2_unpaired.2M.fastq.gz
ILLUMINACLIP:/path/to/Trimmomatic-.039/adapters/TruSeq2-PE.fa:2:30:10 LEADING:3
TRAILING:3 SLIDINGWINDOW:4:30 MINLEN:30
echo "Trimmomatic run complete. Running FastQC on final results..."
```

The script will run more "verbosely" and allow you to track the steps easily.

Galaxy

In this practical, we will show you how to build a program workflow by using the public Galaxy's web server.

(1) Opening Galaxy

Open your desired web browser and go to the Galaxy URL—http://usegalaxy.org/ (Figure 1).

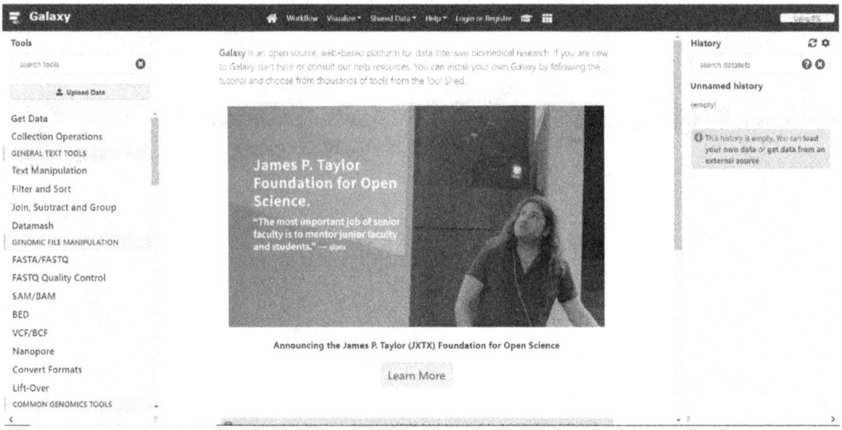

Figure 1. The public Galaxy's web server home page.

(2) Setting up Galaxy account

Go to "Login or Register" page at the top panel to register by entering your email address, password and public name (Figure 2). If you already have an account, you can proceed to login. Once you have registered, you will receive an activation email for verification. Click on the activation link provided and you are able to start now.

Let us take a look at the Analyze Data page (Figure 3).

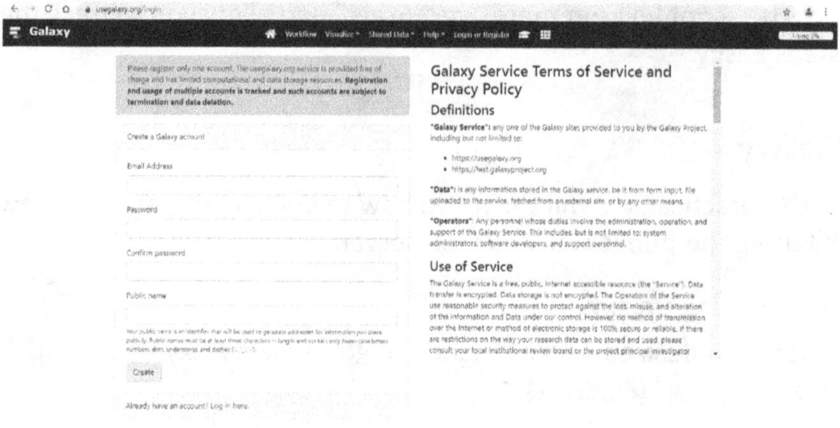

Figure 2. The registration page.

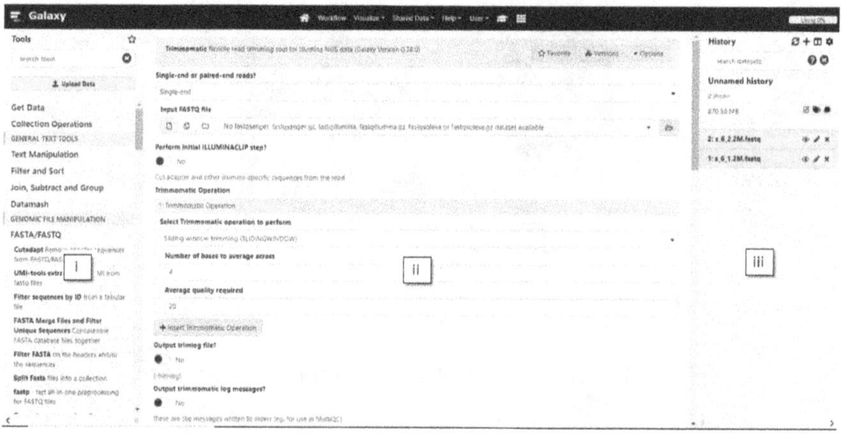

Figure 3. The Analyze Data page.

i. Tool panel — lists of available tools.
ii. Parameter settings panel — allows user to set the conditions to customize the tool. Details of data will also be showing here
iii. History panel — list of jobs executed. Each box represents a job and the status is represented by the following colours:

a. Grey — in queue
b. Yellow — currently running

c. Green — completed successfully
d. Red — failed
e. Light blue — paused

Details of job status can be viewed by expanding the box.

(3) Getting data

The first thing we will do is to get input data. Click "Upload Data" or "Get Data"-> "Upload File from your computer" from the Tool panel. You will see the upload window appearing. Click "Choose local file" and select the 2 input files "s_6_1.2M.fastq" and "s_6_2.2M.fastq". There is no need to worry about zipped and unzipped inputs, as Galaxy can accept both format. Select the data type of input ("fastqsanger" in this example) under **Type** and click "Start", as it will be uncompressed automatically when loaded into history. Change both data type to "fastqsanger" and click "Start". This upload process may take awhile. Click "Close" after files have been uploaded (Figure 4).

You will see 2 boxes representing the 2 datasets appearing in History panel. There are 3 icons at top right of each box in History panel:

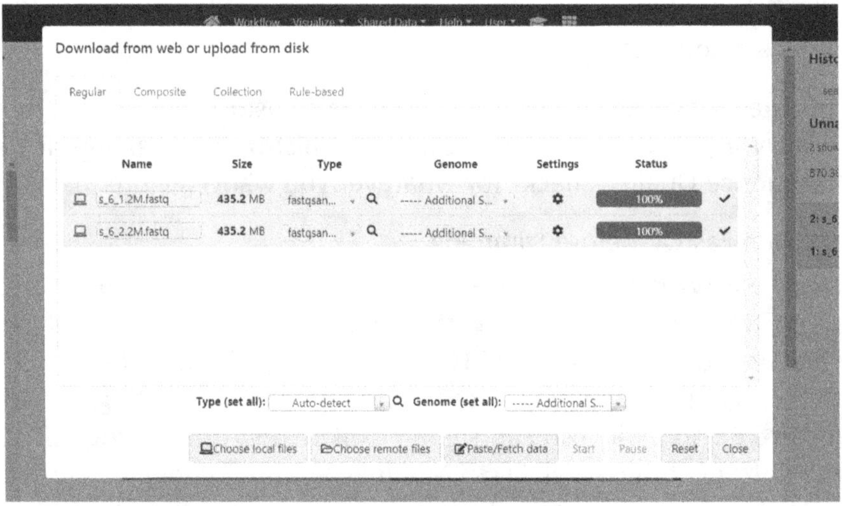

Figure 4. A successful upload.

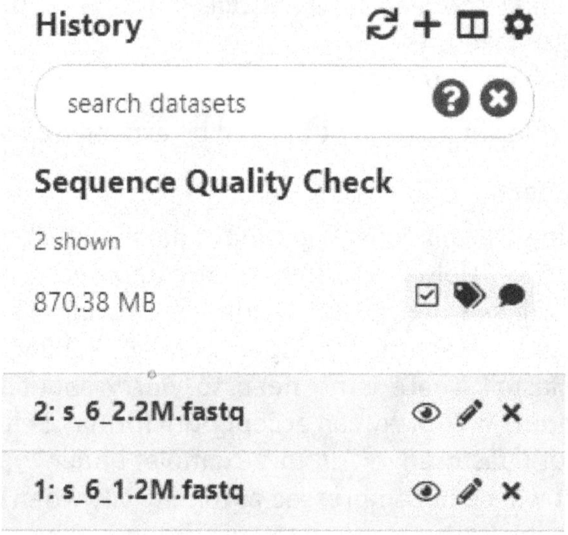

Figure 5. Renamed History; from “Unnamed history” to “Sequence Quality Check”.

i. 👁 - view the datasets
ii. ✏ - edit data details
iii. ✖ - delete data

The setting and steps are tracked as a History. Let us give our History a proper name. Click **Unnamed history** and rename it to “Sequence Quality Check” (or whatever you want) (Figure 5).

(4) Run FastQC before trimming

Next, we want to know the sequence quality of the input files, using the FastQC tool. From the Tool panel, click “FastQC Read Quality reports” under “FASTQ Quality Control”. You should see the program parameters listed in the Parameter settings panel (Figure 6). You will also can see an explanation of the program at the bottom of the panel. Click “Multiple datasets” under **Short read data from your current history**, select the 2 input files, then click “Execute”.

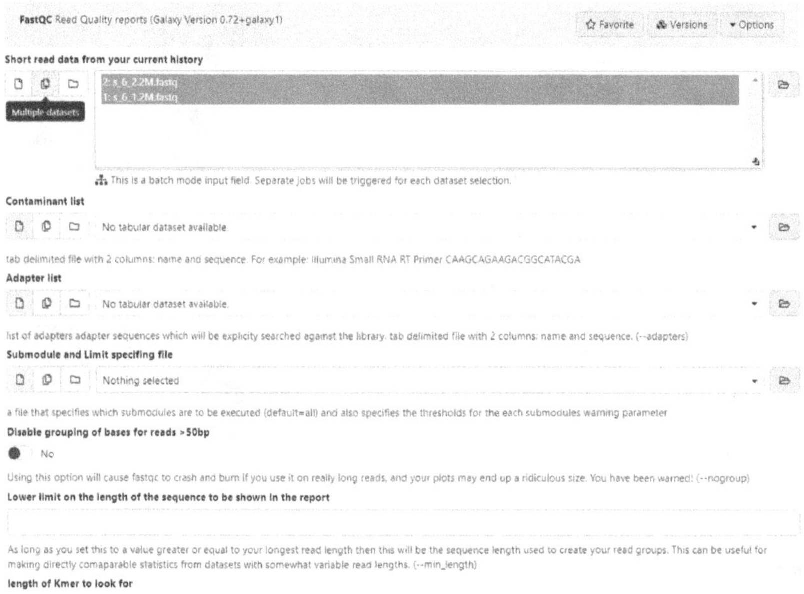

Figure 6. The FastQC parameters in panel.

You will see 4 boxes (representing 4 output files from this tool: "RawData" and "Webpage" format for each input files) added into your History panel, changing from grey to yellow, then becoming green. Click 👁 icon of the box "3: FastQC on data 1: Webpage" to view the result on s_6_1.2M.fastq file (Figure 7).

According to the result, "per base sequence quality" shows that it is unusual; which means the sequence quality is bad. We can see the details of this analysis by clicking the link provided (Figure 8).

The analysis result for another input file s_6_2.2M.fastq showed that the sequence quality begins to fall rapidly after position 60 (Figure 9).

After identifying the quality problems, we now need to trim the sequences that are of low quality. For this, we will use Trimmomatic.

(5) Run Trimmomatic

Click "Trimmomatic" under "FASTQ Quality Control". Select "Paired-end (two separate input files)" for **Single-end or paird-end reads?**, then select the files as **Input FASTQ file R1** and **R2** (Figure 10).

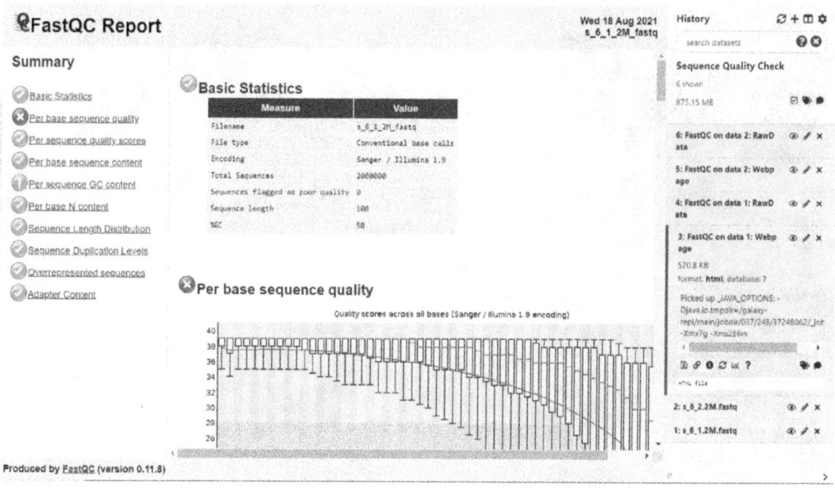

Figure 7. The result of s_6_1.2M.fastq in webpage format.

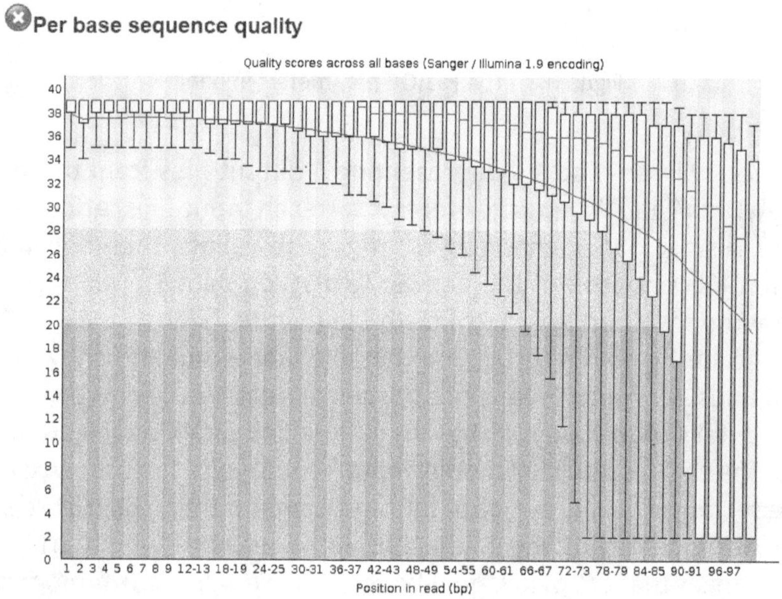

Figure 8. Sequence quality score of s_6_1.2M.fastq file. The quality of the base call dropped starting at position 66.

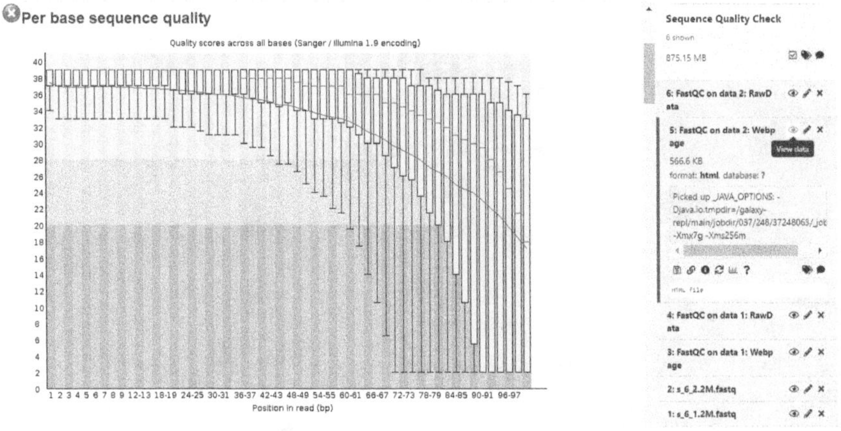

Figure 9. The sequence quality score of s_6_2.2m.fastq file.

Figure 10. The Trimmomatic input in panel.

Figure 11. Example of input files with incorrect data type. Galaxy couldn't detect the files and disabled the selection.

If you encounter a problem stating "No fastqsanger, fastqsanger.gz, ... dataset available" although you have uploaded the input files (Figure 11), this means that the data type of the files were incorrectly set.

You can check the data type of the input files by expanding the boxes in the History panel (Figure 12).

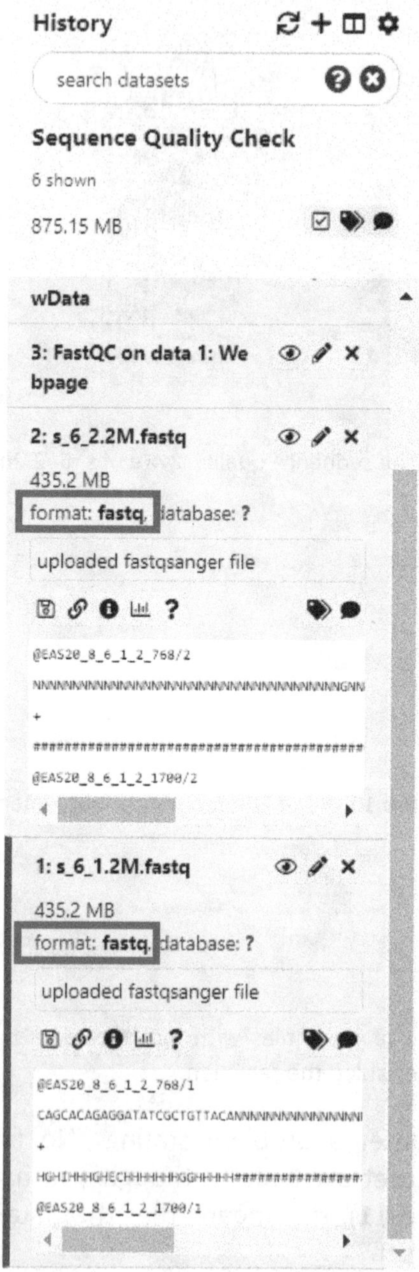

Figure 12. Checking the file format. The format for both datasets is "fastq" instead of "fastqsanger".

To change the data type, click the icon of the box and you will see the Parameter settings panel (Figure 13).

Click the "Datatypes" header, select "fastqsanger" for **New Type** and "Change datatype" (Figures 14 and 15).

Remember to also change the data type for the other dataset.

Once you have changed the data type, you may select the input files for Trimmomatic.

After selecting the input files, make sure you choose "Yes" to **Perform initial ILLUMINACLIP step?**, with the parameters set as shown in Figure 16.

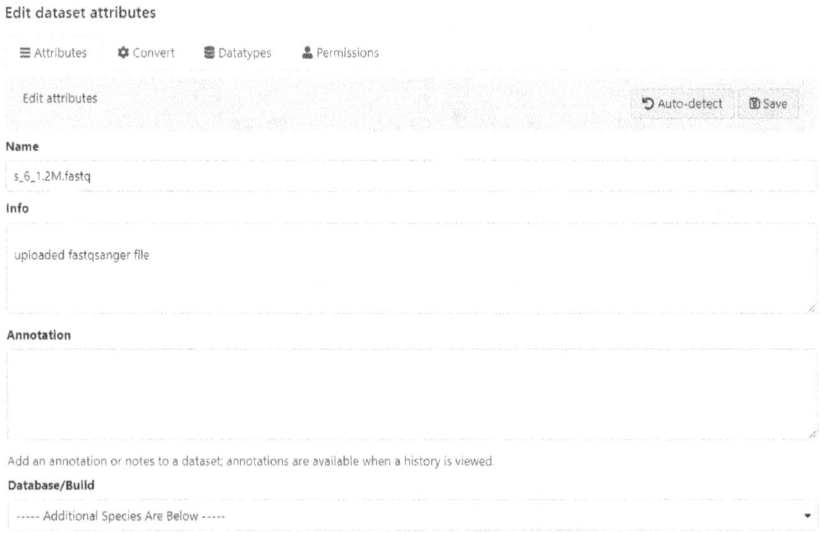

Figure 13. The Parameter settings panel.

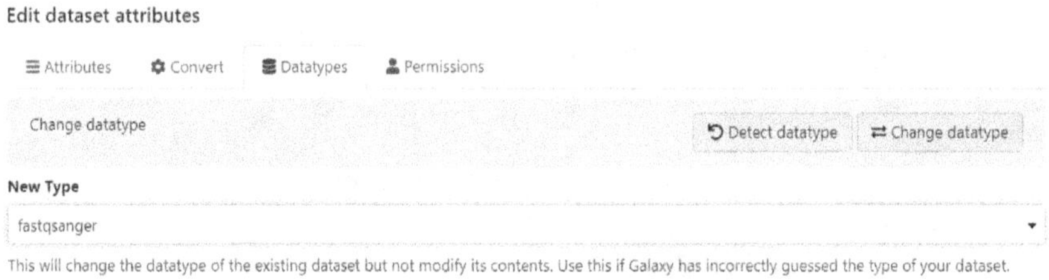

Figure 14. Changing the data type of the file.

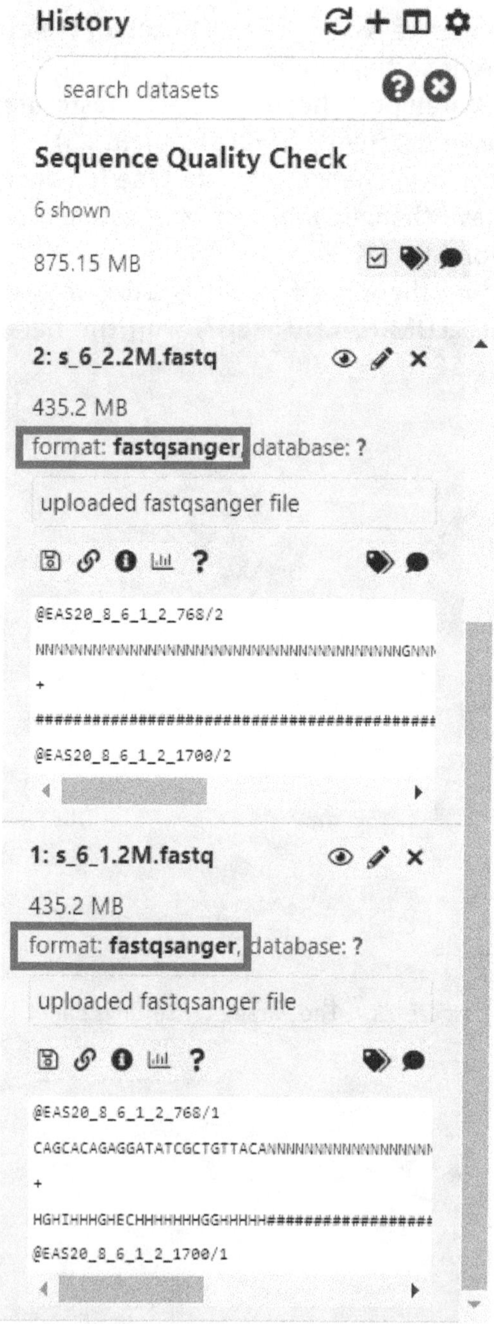

Figure 15. Checking the file format after the change. The format has been changed from “fastq” to “fastqsanger”.

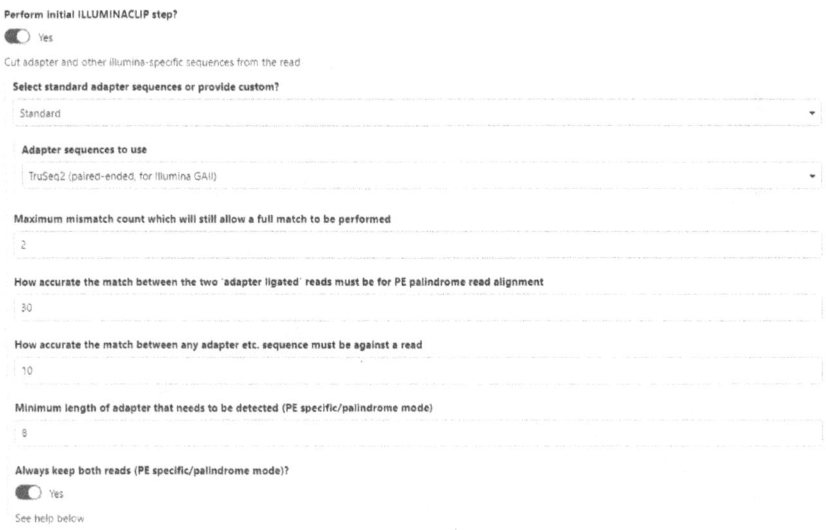

Figure 16. Parameters for performing initial ILLUMINACLIP.

Next, make sure your **Trimmomatic Operation** parameters look exactly as shown in Figure 17, with multiple operations added by clicking "+ Insert Trimmomatic Operation".

Click "Execute". You will see 4 new boxes in your history. The boxes will turn green if the job is completed successfully (Figure 18).

Let us run FastQC again to check if the trimming was effective.

(6) Run FastQC after trimming

Click "FastQC Read Quality reports" under "FASTQ Quality Control". Click "Multiple datasets" under **Short read data from your current history** and select all the output files from the previous step, then click "Execute" (Figure 19).

If you look at the results (Figure 20), you will see the quality of the sequences has improved.

Now, we have completed all steps of our analysis. In Galaxy, we can convert our history into a workflow so that we will be able to execute the same analysis in future.

(7) Create workflow from history

Click the ⚙ button at top of History panel, and select "Extract workflow" (Figure 21).

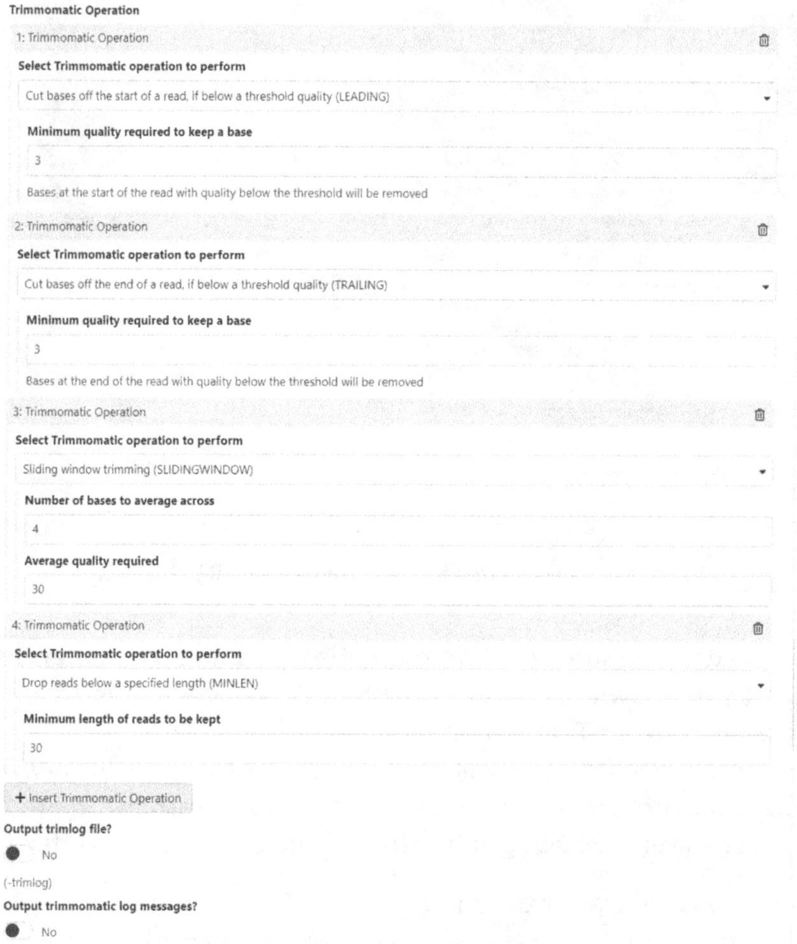

Figure 17. Trimmomatic operations.

Your Parameter settings panel will now look like Figure 22. Select all the steps, and name the workflow as "Sequence Quality Check Workflow" (or whatever you want).

Click "Create Workflow" and you will see the message in Figure 23 appearing.

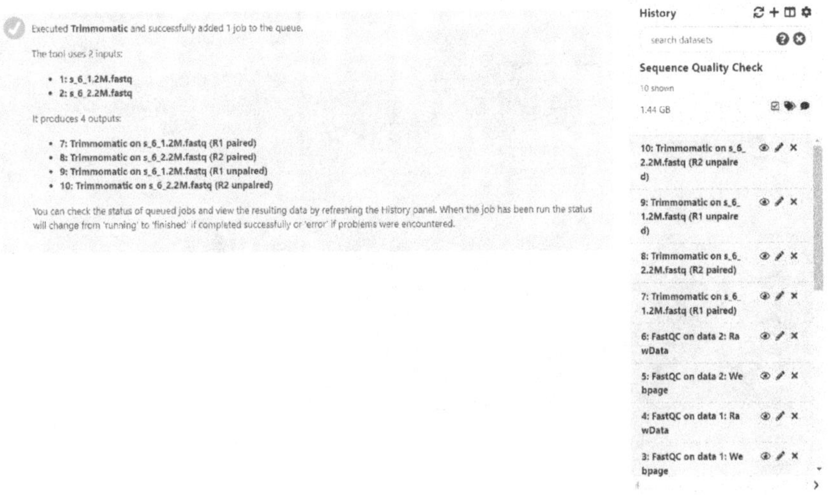

Figure 18. The result after trimming.

Figure 19. The FastQC parameters in the Parameters settings panel.

You can always access your workflows by accessing "Workflow" page on the top panel. Let us make some changes to the workflow, by clicking the workflow name "Sequence Quality Check Workflow" -> "Edit" (Figure 24).

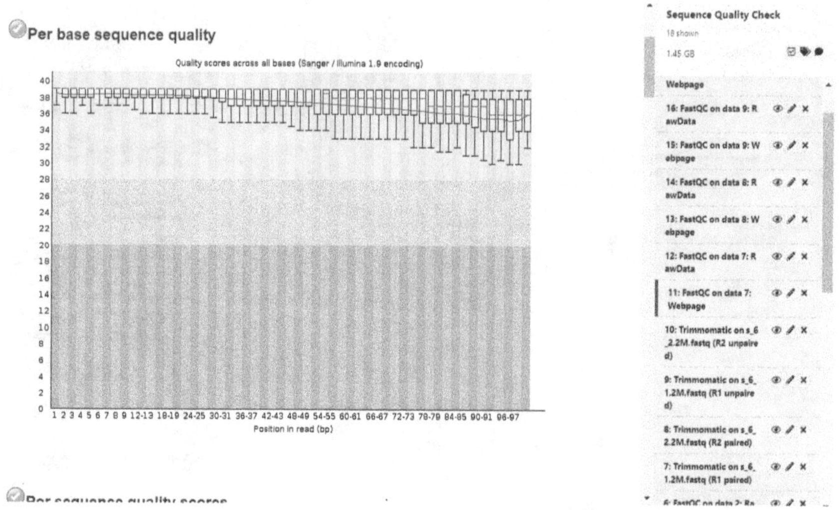

Figure 20. Sequence quality score for s_6_1.2M.fastq file after trimming.

You should see the workflow editor appearing (Figure 25).

i. Tool panel — list of available tools
ii. Editor panel — workflow canvas
iii. Details panel — description of tools, parameter settings
iv. Editor options

Each box in the editor represents a step in the workflow, and the lines connecting the boxes represents the data flow. You may drag and drop the boxes to organize it (Figure 26).

Let us change the name of both input datasets to avoid confusion. Click on the first input dataset "s_6_1.2M.fastq" box and rename it as "R1" in the Details panel (Figure 27). You may also write extra notes under **Step Annotation**. Remember to do the same step for the second input dataset "s_6_2.2M.fastq" but rename it as "R2".

You will notice that there is a checkbox next to every output of each tool. This is used to mark a dataset as the workflow output. By default, all output in a newly created workflow are visible and available. Uncheck an output will make it hidden and not

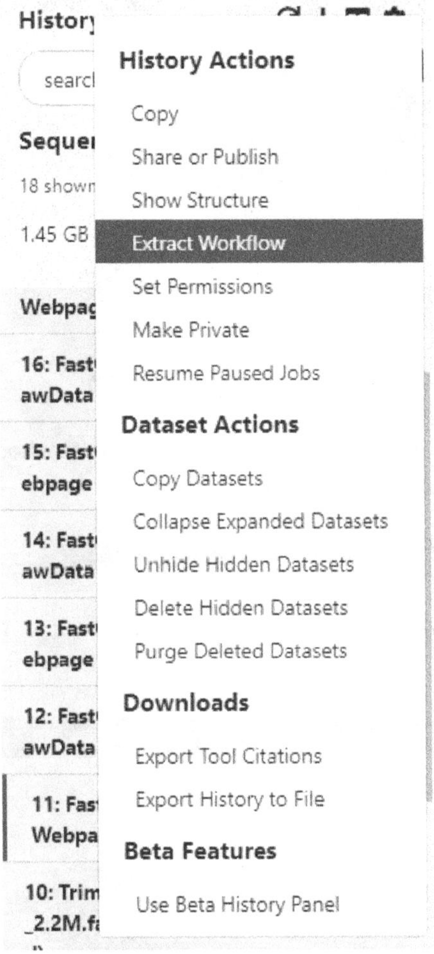

Figure 21. Extracting workflow from history list.

available as the workflow outputs. Let us say if we want the raw data ("RawData (txt)") from FastQC to be hidden, just uncheck the checkbox next to the outputs, as shown in Figure 28.

Take note that the workflow that we built here is a general workflow. From time to time, we are likely to use different parameters when re-running this workflow. Take the parameters in the Trimmomatic software, for example. To enable the parameter change feature, click the "Trimmomatic" box, then in the Details

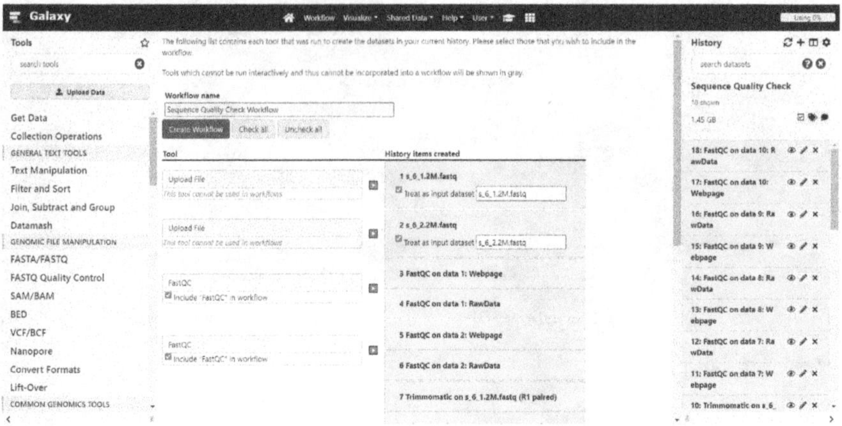

Figure 22. Creating the workflow. You are able to choose which steps to include or exclude from the workflow.

Figure 23. Message appears when the workflow has been created from history.

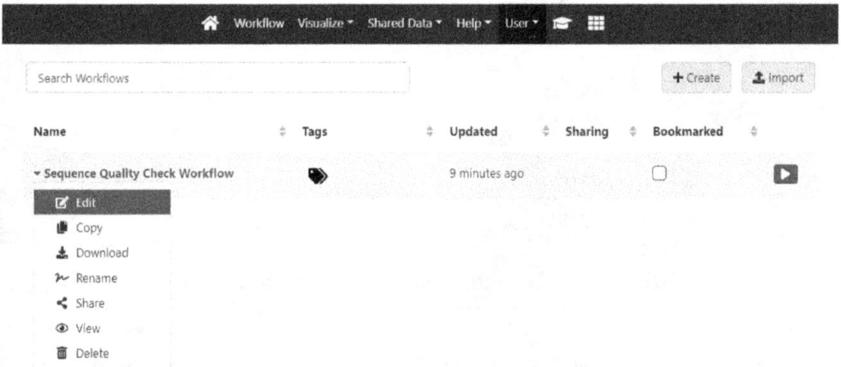

Figure 24. Accessing the workflow editing page.

panel, click the icon of each parameter to change it so that it is pointing down (Figure 29).

When you are done, save the workflow by clicking "Save Workflow" at Editor options.

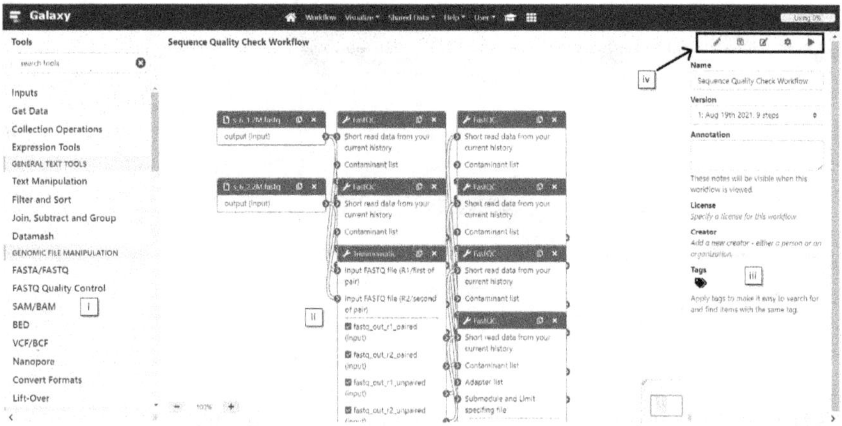

Figure 25. Workflow editor that allows you to edit the workflow settings.

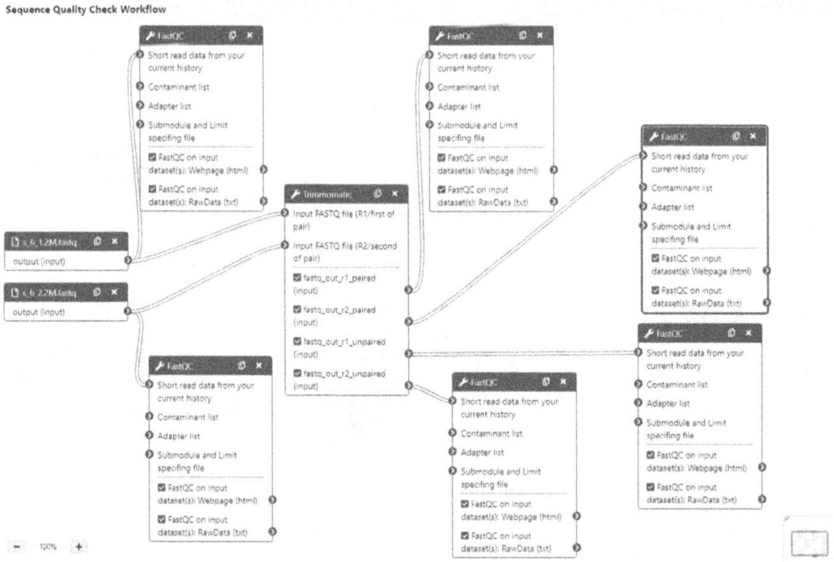

Figure 26. The Editor panel after arrangement.

(8) Run the workflow

Now, let us run our newly created workflow with the same input files again. Click "Run Workflow" icon at the end of "Sequence Quality Check Workflow" line. You can see that all the steps involved in the workflow are listed in the Parameter settings panel.

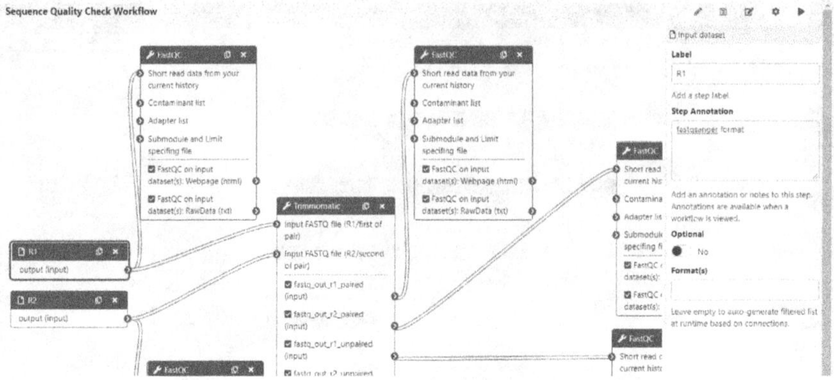

Figure 27. Renaming the steps. The input dataset has been renamed as "R1" in the "Name" attribute under the Details panel, and "fastqsanger format" added to "Step Annotation" attribute.

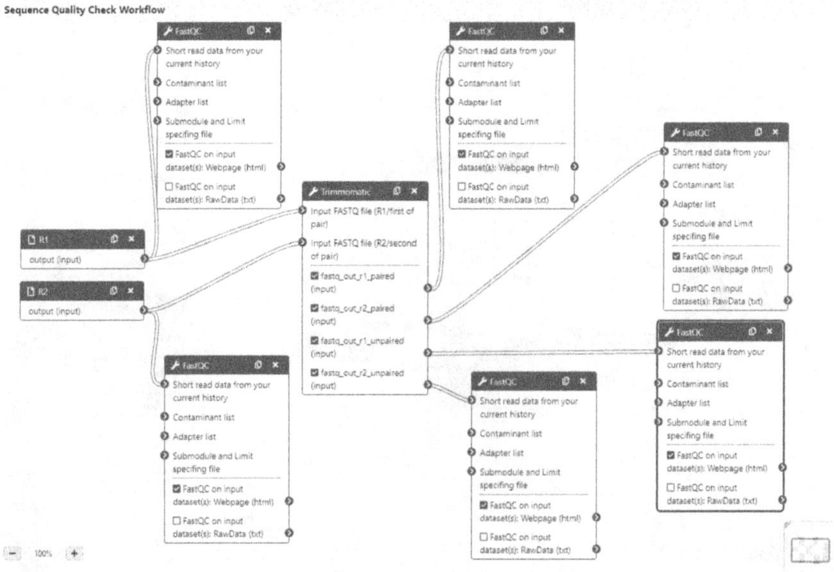

Figure 28. Output settings to determine which file to be hidden or as output from the workflow.

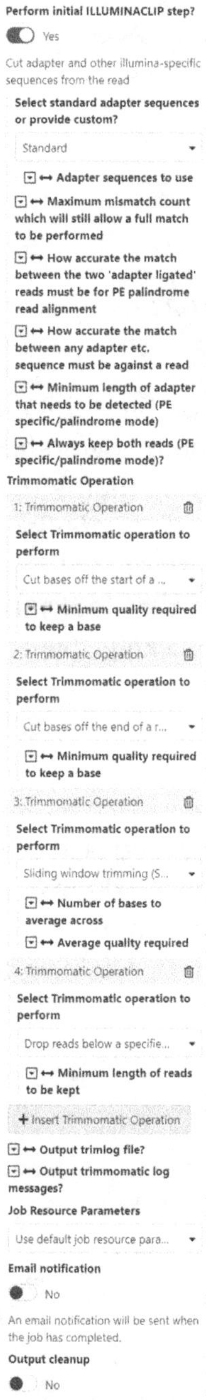

Figure 29. The “Trimmomatic” parameters setting in the Detail panel.

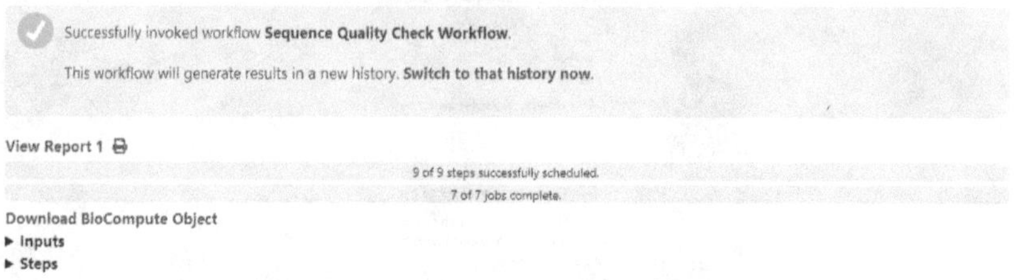

Figure 30. A successful run of the workflow.

Select "s_6_1.2M.fastq" for Step 1:R1 and "s_6_2.2M.fastq" for Step 2:R2 as the input for the workflow. Select "Yes" for **Send results to a new history** and rename the new history. Next, set all the parameters to the same as previous (or whatever you want) for **FastQC** and **Trimmomatic** step. Then, click "Run Workflow". At last, you will see the page that looks like Figure 30. You may see the workflow in detail by clicking "Switch to that history now".

Conclusion

A shell script is a basic means to document and automate a workflow. Galaxy provides a web interface layer that allows a more visually intuitive way to set up workflows compared to shell scripts. Both methods enable researchers to easily revisit or share workflows of their work, as well as retaining the reproducibility of their experiments.

References

1. Bolger, A. M., Lohse, M. & Usadel, B. Trimmomatic: a flexible trimmer for Illumina sequence data. *Bioinformatics* **30**, 2114–2120, doi:10.1093/bioinformatics/btu170 (2014).
2. Andrews, S. *FastQC: a quality control tool for high throughput sequence data*, <http://www.bioinformatics.babraham.ac.uk/projects/fastqc/> (2010).
3. Afgan, E. *et al.* The Galaxy platform for accessible, reproducible and collaborative biomedical analyses: 2018 update, Nucleic Acids Research, Volume 46, Issue W1, 2 July 2018, Pages W537–W544, doi:10.1093/nar/gky379.

Chapter 6

De novo Assembly of a Genome

Joel Low Zi-Bin[a], Martti T. Tammi[a] and Wai Yee Low[b]

[a]Biotechnology & Breeding Department, Sime Darby Plantation R&D Centre, Selangor 43400, Malaysia
[b]The Davies Research Centre, School of Animal and Veterinary Sciences, University of Adelaide, Roseworthy, SA 5371, USA

Glossary of Terms

***De novo*:** Latin expression for "starting from scratch", i.e. without a reference.

Base pair: A pair of complementary nucleotides in the DNA sequence; used as the length unit in genome size measurements.

Read: A string of letters representing DNA sequences generated from sequencing.

Contig: Abbreviation for contiguous sequence.

Scaffold: A "supercontig", containing gaps, which is formed by joining contigs using reads with linkage information between genomic loci, e.g. mate-pair data.

Sequencing depth: The total length of all reads generated by sequencing over the estimated genome's length. It is also called the depth of coverage.

N50: A measure of contiguity of a genome assembly, which refers to the length of the contig/scaffold that is the minimum length required to cumulatively make up 50% of the genome size. The measure can be extended to different genome size percentage cut-offs, such as N75 or N90.

(*Continued*)

(*Continued*)

GC content: Is the percentage of nitrogenous bases in a DNA molecule that are either guanine or cytosine.

Ortholog: Genes in different species that share an evolutionary common ancestor.

De Bruijn assembler: An assembler that models the relationship between exact k-mers from the reads. The nodes in the graph represent k-mers, and the edges represent the overlap of adjacent k-mers by k − 1 letters. Assembly is done by tracing the path with most consistency through the graph.

Overlap layout consensus (OLC) assembler: An assembler that identifies all pairs of reads that overlap sufficiently well and then organizes this information into a graph containing a node for every read and an edge between any pair of reads that overlap each other. Contigs are generated as a consensus by inferences from information of all edges in the possible path.

Introduction

The ***de novo*** assembly of a genome is quite an art as it is an attempt to build a finished product without knowing how it actually looks like. Many of the current sequencing technologies employ methods to first size select the DNA fragments prior to sequencing, and thus, the entire length of an organism's DNA is not read in a single run. Even on a PacBio Sequel, which is an established long **read** third-generation sequencer, the average read is 10,000-14,000 bp,[1,2] while the smallest bacteria genome[3] currently known is still 160,000 bp long.

Current sequencing processes require the fragmentation of the genome for sequencers to read. A collection of fragments is called a library and it is usually categorized according to its fragment lengths (e.g. 20 kb, 3 kb, or 100 bp libraries). In addition, libraries can also be categorized according to the methods used to sequence the fragments. Taking the Illumina platform as an example, a genome is typically fragmented into 300 bp long pieces that get inserted between adaptors. The sequencing of the insert from the two ends of the fragments creates a pair of reads. This is known as a 300 bp paired-end (PE) read library. The sequenced reads are shorter than the entire fragment (e.g. 75 bp). Singletons, also known as single-end (SE) or orphaned reads, consist of reads that are sequenced only from one end of the DNA insert. There is another pairing technique that is used to create libraries that span even greater distances between 1 kbp and 150 kbp. These are called mate-pair libraries[4] and are achieved by circularizing the large inserts with the ends marked and joined together to be sequenced.

Once the reads are generated from these libraries and trimmed for the best quality, software called assemblers are used to assemble them, much like putting together a jigsaw puzzle. The end results are contigs. A **contig** is a contiguous length of genomic sequence in which the order of bases is known to have a high confidence level. However, it is possible that chimeric contigs are generated in the assembly process,[5] which may be detected using the optical map. A **scaffold** is a portion of the genome reconstructed from contigs and contains gaps (Figure 1). Gaps occur where information between

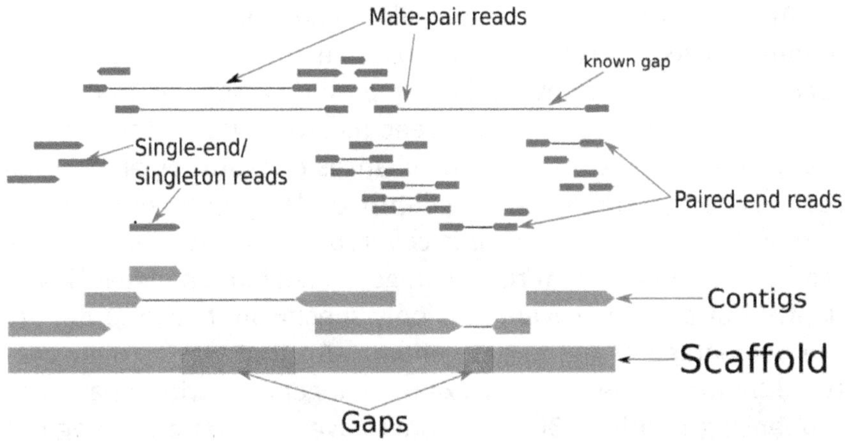

Figure 1. An illustration of the assembly of singleton, paired-end, and mate-pair reads to form a scaffold.

contigs are unavailable. Possible causes for gaps are sequence repeats or unsequenced regions of the genome. The unsequenced parts of the insert for PE and mate-pair reads become gaps in a scaffold when no overlapping sequence can be found for the region. While mate-pair libraries have been used to scaffold contigs, nowadays it is more common to use HiC-seq that is able to link reads up to 10,000 kb apart in the genome.[6]

The puzzle is rarely complete. A great deal of experience and knowledge on both the organism's genome and the tools used is needed to get the optimal, rarely perfect, results. The puzzle is harder to solve if the genome in question is complex. In general, the complexity of a genome increases as it gets larger, contains higher **GC content**, has lots of repeats, and/or contains a higher number of chromosomes.

One of the biggest challenges in sequencing and the cause of many gaps in an assembly is the presence of many repeats in the genome. Repeats usually occur in tandem, not necessarily identical, and stretch to very long lengths. Assemblers identify repeats when reads are sectioned into parts of fixed-length strings. An index of all possible combinations can be made for a particular string length (k) and searched to see how repeated the pattern is in the genome. The k-mer refers to this index.[7,8]

An important determinant of a complete genome is the amount of its entirety that one is able to sequence. Sufficient reads are needed to cover all the gaps in an assembly. To do that, it may be required to sequence deep (i.e. to sequence the genome many times over to increase the chances of capturing all possible read overlaps). The **sequencing depth**, or depth of coverage, is the number of times a sequence is covered by the total length of all reads. Sometimes, the term coverage is used interchangeably with depth[9] but *genome coverage* is only specific in its use to denote the percentage of the target genome size that was actually captured by the sequences. For example, a genome with an average sequencing depth of 30× may only have a genome coverage of 95% (Figure 2).

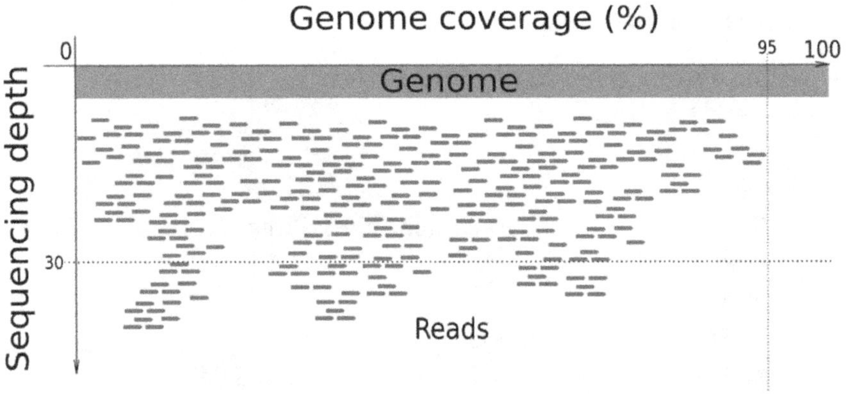

Figure 2. Sequencing depth vs. genome coverage.

It is important to understand that different assemblers use different algorithms for specific sequencing libraries. It is best to understand the strengths and weaknesses of each assembler and use the optimal ones for your types of libraries or purposes. One may need to refine assembly parameters and conduct multiple assemblies to obtain the desired results. Three reviews on genome sequence assembly were written in 2013,[4] 2015,[10] and 2017[11] that would be good for further reading on the theoretical basis of current sequencing methods and best practices.

In the following practical, we will assemble a bacteria genome, *Escherichia coli (E. coli),* using the latest tools as of this writing. We will assemble three genome drafts from two types of libraries and then compare the quality of each assembly.

Overall Steps

(1) Download sequences
(2) Filter out bad reads
(3) Assemble the genome(s)
(4) Check the quality of the genome(s)

Download Sequences

- Set up a working directory to put all files:
  ```
  $ mkdir chapter6_runs
  $ cd chapter6_runs
  ```
- Illumina PE (short):
  - Source: https://trace.ncbi.nlm.nih.gov/Traces/sra/?run=ERR008613.
  - Info:
    - Files are *gz* compressed
    - Number of reads: 14,214,324 ∗ 2 = 28,428,648
    - Read length (bp): 2 × 100 (note that 2× means this is a paired end read)
    - Insert size (bp): 215.4 ± 10.6
  - Download the two files:
  - http://ftp.sra.ebi.ac.uk/vol1/run/ERR008/ERR008613/200x100x100-081224_EAS20_0008_FC30TBBAAXX-6.tar.gz.
- PacBio SE (long):
  - Source: https://github.com/PacificBiosciences/DevNet/wiki/E.-coli-Bacterial-Assembly.
  - Note that we used the uncompressed final output (polished_assembly.fastq.gz) as a genome reference later in Step 4.
  - Info:
    - Instrument: PacBio RS II
    - Chemistry: C4
    - Enzyme: P6
    - One SMRT Cell
    - Size selected 20kb library
    - Number of reads: 13,124
  - The raw data are downloadable from: https://s3.amazonaws.com/files.pacb.com/datasets/secondary-analysis/e-coli-k12-P6C4/p6c4_ecoli_RSII_DDR2_with_15kb_cut_E01_1.tar.gz.
  - The raw data generated by the PacBio sequencer are usually processed through the SMRT Analysis suite that comes with the machine to produce a fastq file that is filtered of the SMRT system

adapters. This fastq file is what we will use and is available here (from the CANU tutorial later explained below): http://gembox.cbcb.umd.edu/mhap/raw/ecoli_p6_25x.filtered.fastq. Rename this file to pacbio.fastq.

- Oxford Nanopore (long):
  - The data were released by Loman in http://lab.loman.net/2015/09/24/first-sqk-map-006-experiment/
  - http://nanopore.s3.climb.ac.uk/MAP006-PCR-1_2D_pass.fasta.

Filter Out Bad Reads

Use: **Trimmomatic**[12](http://www.usadellab.org/cms/?page=trimmomatic.)

Key features:

(1) Takes a file of multiple sequences to match against the reads for removal. Mainly used to remove sequencing adapters but can be used for contaminant removal as well.
(2) Reads and outputs compressed fastq files for the storage conscious.
(3) Keeps orphaned pairs to be used as SE reads.

Version used: V0.32
Installation: Download the binary version from the website and unzip the file. You will need Java installed to run the trimmomatic-0.38.jar file.

```
$ wget http://www.usadellab.org/cms/uploads/supplementary/Trimmomatic/Trimmomatic-0.38.zip
$ unzip Trimmomatic-0.38.zip
```

The command to run:

```
$ java -jar Trimmomatic-0.38/trimmomatic-0.38.jar PE ../raw_data/EAS20_8/s_6_1.fastq ../raw_data/EAS20_8/s_6_2.fastq s_6_1_paired.fastq.gz s_6_1_unpaired.fastq.gz s_6_2_paired.fastq.gz s_6_2_unpaired.
```

```
fastq.gz ILLUMINACLIP:/apps/software/Trimmomatic/0.38-
Java-1.8.0_121/adapters/TruSeq2-PE.fa:2:30:10 LEADING:
3 TRAILING:3 SLIDINGWINDOW:4:30 MINLEN:30
```

Removes Illumina adapters given in TruSeq2-PE.fa (provided by Trimmomatic in the "adapters" directory)

- Remove leading and trailing edges of reads with low quality or N bases (below quality 3).
- Scan the read with a 4-base wide sliding window, cutting when the average quality per base drops below 30 (SLIDINGWINDOW:4:30).
- Removes reads that are shorter than 30 bases.
- Note that the raw Illumina fastq files were downloaded to a folder named "raw_data". The "../raw_data" means that this folder is just one level above in the directory structure from the current directory where we run the trimmomatic command.

For more information about read quality and trimming, please refer to Chapter 3.

Important Output file(s):

(1) s_6_1_paired.fastq.gz
(2) s_6_1_unpaired.fastq.gz
(3) s_6_2_paired.fastq.gz
(4) s_6_2_unpaired.fastq.gz

Runtime: The run took 20 min on an Intel(R) Xeon(R) Gold 6248 CPU @ 2.50GHz machine. In this run, 2 CPU cores were given and in total 0.69 core-hours were used and less than 4 GB RAM was needed. The following examples are based on the same system.

Summary results:
Input Read Pairs: 14,214,324; Both Surviving: 10,707,272 (75.33%); Forward Only Surviving: 1,525,592 (10.73%); Reverse Only Surviving: 1,175,972 (8.27%); Dropped: 805,488 (5.67%).

Assemble the Genome(s)

Short paired-end read assembly

Use: **SPAdes**[13] (https://github.com/ablab/spades.)
Key features:

(1) Current best for simple/small/microbe genomes.
(2) **De Bruijn assembler** optimized for short reads.
(3) Supports many sequencing platforms' outputs e.g. Illumina, PacBio.

Version used: 3.12.0
Installation: Download the Linux binaries version from the website and unpack the files. You will also need Python installed (comes with any Linux OS).

```
$ wget http://cab.spbu.ru/files/release3.12.0/SPAdes-3.12.0-Linux.tar.gz
$ tar -xzf SPAdes-3.12.0-Linux.tar.gz.
```

The command to run:

```
$ spades.py --pe1-1 s_6_1_paired.fastq.gz --pe1-2 s_6_2_paired.fastq.gz --pe1-s s_6_1_unpaired.fastq.gz --pe1-s s_6_2_unpaired.fastq.gz --careful -t 4 -o ecoli_illupe
```

- Reads that are still paired after the Trimmomatic run are identified with the --pe argument. The first number after "pe" refers to the arbitrary library number the reads originate from, while the subsequent number refers to the pairing, i.e. 1 for forward, 2 for reverse.
- Reads that did not survive pairing after the Trimmomatic run are identified with the same --pe argument with the addition of the –s flag.
- We added the --careful argument to reduce mismatches and short indels at the expense of a longer run time.

- The -t argument is optional and in this case, we have asked for 4 threads.
- The output is gathered in a directory named in the –o argument.

Important Output file(s):

(1) ecoli_illupe/scaffolds.fasta -> rename to ecoli_illupe.fasta
Runtime: 64 min. Given 4 cores and 16 GB memory.
Summary results: 277 scaffolds.

Hybrid assembly with PacBio reads

Use: **SPAdes** (https://github.com/ablab/spades.)
Key features and installation: (refer to "Short paired-end read assembly" above).

Both Illumina and PacBio reads belong to the same *E. coli* strain, K-12 MG1655. You should only assemble reads from the same organism. Otherwise, the results of the assembly may be of poor quality.

PacBio reads are excellent for gap closure and repeat resolution because the average read length of this platform is long (e.g. >10 kb). There are different categories of reads from this platform and for the purpose here, use filtered subreads in FASTQ/FASTA format.

The command to run:

```
$ python SPAdes-3.6.1-Linux/bin/spades.py --pe1-1
s_6_1_paired.fastq.gz --pe1-2 s_6_2_paired.fastq.
gz --pe1-s s_6_1_unpaired.fastq.gz --pe1-s s_6_2_
unpaired.fastq.gz --pacbio pacbio.fastq --careful
-o ecoli_illupe-pacbio
```

- An important parameter to adjust: -k, which determines the k-mer size to index.
- Spades does not assemble PacBio reads directly. It uses such long reads as a scaffold to improve the contiguousness of the assembly when combined with short reads sequences. The argument --pacbio is used here to refer to the PacBio reads in fastq format.

Important output file(s):

(1) illupe-pacbio/scaffolds.fasta -> rename to ecoli_illupe-pacbio.fasta

Runtime: 78 min. Given 4 cores and 16 GB memory.
Summary results: 42 scaffolds.

Long SE read assembly (PacBio)

We will use the Canu[14] assembler (https://github.com/marbl/canu/releases/tag/v2.0), which is based on the **Celera Assembler**[15] (http://wgs-assembler.sourceforge.net/.)

Key features:

(1) **Overlap Layout Consensus (OLC)** assembler optimized for long reads.
(2) Supports all long SE read (no shorter than 75 bases) platforms, i.e. Sanger, 454, Illumina, PacBio, Oxford Nanopore.
(3) Established pipelines.

Long reads use a different algorithm for assembly, which is called the OLC method. One of the oldest and still widely used programs that use this algorithm is the Celera Assembler.

Celera Assembler is a part of the SMRT Analysis suite of programs that is used to clean, process, and assemble PacBio reads. There is a pipeline called CANU that contains steps to assemble the *E. coli* genome: http://canu.readthedocs.org/en/latest/quick-start.html#quickstart.

Version used: CANU 2.0

Installation: Download the Linux binaries version from the website and unpack the file.

The command to run:

```
$ canu -p ecoli -d ecoli-pacbio genomeSize=4.8m
corThreads=2 gridOptions="-N 1" -pacbio pacbio.
fastq
```

- CANU is told where the PacBio fastq file is with the –pacbio argument.
- The output is gathered in the directory named using –d with files having prefixes named using –p.
- The genomeSize parameter can be a rough estimate.
- Note that the gridOptions="-N 1" argument is probably not necessary for you but it is required on the server where this was run.

Important Output file(s):

(1) ecoli.contigs.fasta
Rename to ecoli_pacbio.fasta.
Runtime: 1 h 10 min.
Summary results: 2 scaffolds.

Long SE read assembly (Oxford Nanopore)

The command to run:

```
$ canu -p ecoli -d ecoli-oxford genomeSize=4.8m
corThreads=2 gridOptions="-N 1" -nanopore oxford.
fasta
```

- CANU is told where the Oxford Nanopore fasta file is with the –nanopore argument.
- Other arguments are similar to that explained in the PacBio section above.

Important Output file(s):

(2) ecoli.contigs.fasta
Rename to ecoli_oxford.fasta
Runtime: 2 h 24 min (note this run time is longer due to waiting for resources to be available on the server).
Summary results: 4 scaffolds.

Check the Quality of the Genome

There are essentially two metrices to assess the quality of the genome:

(1) Statistical
(2) Evolutionary

The four assemblies are assessed as follows:

Statistical

Use: **QUAST**[16] (http://quast.sourceforge.net/quast.html.)
Key features:

(1) Works both with and without a given genome reference.
(2) Able to do multiple genome comparisons.
(3) Generates interactive reports that can be opened in web browsers.

Version used: 4.5
Installation: Download the source code from the website and unpack the file. You will also need python installed (comes with any Linux OS). QUAST installs the necessary software on the fly too during its first use (see the README.md).

Since we are doing *de novo* assembly we assume that a reference genome is not available. We will proceed to compare the metrices:

```
$ quast.py ecoli_illupe.fasta ecoli_illupe-pacbio.
fasta ecoli_pacbio.fasta ecoli_oxford.fasta --
glimmer --scaffolds -o ecoli_quast
```

Alternatively, if there is a genome reference, we can use additional options (-R) to give us a better picture of the assembly quality.

```
$ quast.py ecoli_illupe.fasta ecoli_illupe-pacbio.
fasta ecoli_pacbio.fasta ecoli_oxford.fasta --
glimmer --scaffolds -R polished_assembly.fasta -o
ecoli_quast_withref
```

- The polished_assembly.fasta is the file downloaded in Step 1.
- The arguments are self-explanatory. Run quast.py without any arguments to see details for all arguments.

- Read more on the following parameters because they are likely useful for your analysis: --gage, --contig-thresholds, --use-all-alignments, --ambiguity-usage, --strict-NA, --extensive-mis-size

Important Output file(s):

(1) ecoli_quast/report.pdf
(2) ecoli_quast/report.html
(3) ecoli_quast_withref/report.pdf
(4) ecoli_quast_withref/report.html

The html files can be opened in a web browser for interactive reports.

Runtime: less than 5 min.

Figures 3 and 4 provide examples of results.

19 April 2021, Monday, 13:55:38

View in Icarus contig browser

All statistics are based on contigs of size >= 500 bp, unless otherwise noted (e.g., "# contigs (>= 0 bp)" and "Total length (>= 0 bp)" include all contigs).

Worst Median Best ☑ Show heatmap

Statistics without reference	ecoli_illupe	ecoli_illupe_broken	ecoli_illupe-pacbio	ecoli_illupe-pacbio_broken	ecoli_pacbio	ecoli_oxford
# contigs	86	92	5	7	2	4
# contigs (>= 0 bp)	277	92	42	7	2	4
# contigs (>= 1000 bp)	74	80	1	2	2	4
# contigs (>= 5000 bp)	57	62	1	2	2	3
# contigs (>= 10000 bp)	53	58	1	2	2	3
# contigs (>= 25000 bp)	46	47	1	2	1	2
# contigs (>= 50000 bp)	32	32	1	2	1	2
Largest contig	265 168	265 168	4 641 533	4 305 765	4 655 056	2 479 943
Total length	4 556 193	4 555 615	4 643 994	4 642 993	4 673 397	4 641 622
Total length (>= 0 bp)	4 582 658	4 555 615	4 649 534	4 642 993	4 673 397	4 641 622
Total length (>= 1000 bp)	4 547 585	4 547 007	4 641 533	4 639 991	4 673 397	4 641 622
Total length (>= 5000 bp)	4 506 454	4 505 969	4 641 533	4 639 991	4 673 397	4 637 759
Total length (>= 10000 bp)	4 477 091	4 476 606	4 641 533	4 639 991	4 673 397	4 637 759
Total length (>= 25000 bp)	4 373 933	4 306 277	4 641 533	4 639 991	4 655 056	4 622 121
Total length (>= 50000 bp)	3 879 465	3 769 964	4 641 533	4 639 991	4 655 056	4 622 121
N50	133 458	133 019	4 641 533	4 305 765	4 655 056	2 479 943
N75	64 855	61 158	4 641 533	4 305 765	4 655 056	2 142 178
L50	13	13	1	1	1	1
L75	25	26	1	1	1	2
GC (%)	50.74	50.74	50.79	50.79	50.73	51
Mismatches						
# N's	578	0	1001	0	0	0
# N's per 100 kbp	12.69	0	21.55	0	0	0
Predicted genes						
# predicted genes (unique)	3594	3595	3574	3602	4166	5029
# predicted genes (>= 0 bp)	3594	3595	3608	3636	4185	5047
# predicted genes (>= 300 bp)	3376	3377	3385	3410	3769	3932
# predicted genes (>= 1500 bp)	661	661	663	667	477	205
# predicted genes (>= 3000 bp)	85	85	87	87	46	8

Figure 3. HTML Extended Report from QUAST results without a reference genome.

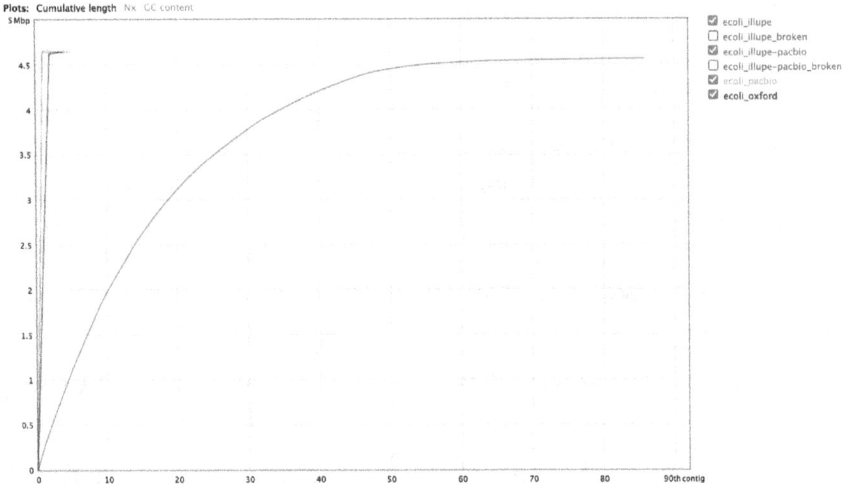

Figure 4. Plot of cumulative lengths in the interactive QUAST reports.

The QUAST results show that the Pacbio assembly is the best because it has only two contigs and scaffold N50 of 4,655,056 bp that matches the total length of the genome.

Evolutionary

Use: **BUSCO**[17] (http://busco.ezlab.org/.)
Key features:

(1) Uses the biological basis of universal single-copy **orthologs** to benchmark the genome quality.
(2) A successor to CEGMA (http://korflab.ucdavis.edu/datasets/cegma/.)

Version used: v2.0.1

Installation: Download the program and unpack it. There are a number of dependencies needed to get BUSCO running. Do refer to the BUSCO_userguide.pdf file for detailed instructions. The following dependencies were used:

(a) BLAST+ 2.6.0
(b) hmmer-3.1b2
(c) augustus-3.2.3

BUSCO searches for the presence of evolutionary conserved genes. Single-copy orthologs from a clade or group of species related to the newly decoded genome are used for comparison. Download the necessary BUSCO profile for your type of organism. In our case, we use the bacteria profile which can be downloaded:

```
wget http://busco.ezlab.org/v2/datasets/bacteria_
odb9.tar.gz.
```

The file should be unpacked in a location, such as BUSCO's directory (/path/to/BUSCO_v2.0.1/) or the path where you have the assembly fasta files:

```
$ tar -zxvf bacteria_odb9.tar.gz
```

Make the runs for each assembly:

(1) Illumina only assembly:

```
$ BUSCO.py -i ecoli_illupe.fasta -o ecoli_illupe
-c 4 -l bacteria_odb9/ -m geno
```

(2) PacBio only assembly:

```
$ BUSCO.py -i ecoli_pacbio.fasta -o ecoli_pacbio
-c 4 -l bacteria_odb9/ -m geno
```

(3) Illumina + PacBio assembly:

```
$ BUSCO.py -i ecoli_illupe-pacbio.fasta -o ecoli_
illupe-pacbio -c 4 -l bacteria_odb9/ -m geno
```

(4) Oxford Nanopore assembly:

```
$ BUSCO.py -i ecoli_oxford.fasta -o ecoli_oxford
-c 4 -l bacteria_odb9/ -m geno
```

Important Output file(s):
/run_ecoli_illupe/short_summary_ecoli_illupe.txt

Table 1. BUSCO results.

	Illumina	Illumina+PacBio	PacBio	Oxford
Complete single-copy BUSCOs	146	146	116	44
Complete duplicated BUSCOs	0	0	0	0
Fragmented BUSCOs	0	0	17	55
Missing BUSCOs	2	2	15	49
Total BUSCO groups searched	148	148	148	148

/run_ecoli_pacbio/short_summary_ecoli_pacbio.txt
/run_ecoli_illupe-pacbio/short_summary_ecoli_illupe-pacbio.txt
/run_ecoli_oxford/short_summary_ecoli_oxford.txt

Runtime: 4 min.
Summary results in Table 1.

The BUSCO results indicate that both Illumina and hybrid Illumina + PacBio assemblies are the best because they only missed 2 known BUSCO genes in 148 bacterial genes. The Oxford Nanopore assembly is the worst according to the BUSCO metric as it has the highest number of fragmented and missing BUSCO genes.

Note:

(1) Ensure the right versions of the dependencies are used (e.g. version 3.0.3 for augustus).
(2) Ensure the dependencies can be called in the environment by setting the paths.

Discussion and Conclusion

To ensure that only good quality data are used for assembly, filtering of the raw FASTQ reads after a sequencing run is necessary. Besides filtering out sequencing adapter sequences, one could look at implementing sequence filters for contaminant sequences. In addition, one should be mindful of the effective sequencing depth after trimming away unwanted sequences. Generally, a 30× effective sequencing depth is considered good for Illumina sequenced data.

There are many assemblers available for use. The use of hybrid assemblers such as SPAdes gives users more options on the choices of input sequences and this may improve the chances of getting a correctly assembled genome when compared to assemblers that only accept a single type of input. The ability to use multiple inputs allows users to combine the strength of long reads from platforms such as the PacBio with the advantages of highly accurate short reads from the Illumina.

PacBio-based assembly of the *E. coli* genome is the best according to the statistical assessment by QUAST (e.g. higher contig N50 and 1 major contig that contains the equivalent of the expected genome size). Although the QUAST results of the PacBio-based assembly appeared superior, it only painted one side of the story.

The use of an evolutionary basis in assessing genome drafts is crucial as one of the objectives of genome science is to analyse genes, which requires accurate sequences to aid genome annotation. Programs like BUSCO and CEGMA[18] allow users to check if the draft assembly contains highly conserved genes that tend to occur as single copy among organisms that belonged to a particular clade. In our example, both the Illumina and hybrid assemblies showed the best results in two ways; firstly by having the most number of intact single-copy ortholog genes for bacteria and secondly by having no duplicated or fragmented ortholog. However, it should be noted that the read coverage in PacBio assembly was only ~24×, which is less than ideal. If ~30× PacBio coverage was used, the BUSCO result may be substantially better. The Oxford Nanopore read coverage was ~29× and hence, it is clear that the early sequencing chemistry based on this platform is more erroneous.

The assembly process for even a bacterium genome is not trivial. Our example did not produce the optimal assembly in its first run. A few iterations are usually needed before a satisfactory result is obtained. Methods to improve the assembly results include filtering both Illumina and PacBio reads further, tweaking SPAdes parameters, and adding a subsequent genome polishing step. It may be viable to assemble a PacBio-only draft first with more stringent overlap cut-offs, which are later used in SPAdes. Furthermore, since the *E. coli* genome is circular, we should further identify overlapping ends of the final assembled contig and trim it with programs such as Circlator.[19]

References

1. Buermans, H. P. J. & den Dunnen, J. T. Next generation sequencing technology: Advances and applications. *Biochimica et Biophysica Acta — Molecular Basis of Disease* **1842**, 1932–1941 (2014).
2. Ardui, S., Ameur, A., Vermeesch, J. R. & Hestand, M. S. Single molecule real-time (SMRT) sequencing comes of age: Applications and utilities for medical diagnostics. *Nucleic Acids Research* **46**, 2159–2168 (2018).
3. Nakabachi, A. *et al.* The 160-kilobase genome of the bacterial endosymbiont Carsonella. *Science* **314**, 267 (2006).
4. Nagarajan, N. & Pop, M. Sequence assembly demystified. *Nature Reviews Genetics* **14**, 157–167 (2013).
5. Low, W. Y. *et al.* Haplotype-resolved genomes provide insights into structural variation and gene content in Angus and Brahman cattle. *Nature Communications* **11**, 1–14 (2020).
6. Lieberman-Aiden, E. *et al.* Comprehensive mapping of long-range interactions reveals folding principles of the human genome. *Science* **326**, 289–293 (2009).
7. Kurtz, S., Narechania, A., Stein, J. C. & Ware, D. A new method to compute K-mer frequencies and its application to annotate large repetitive plant genomes. *BMC Genomics* **9**, 517 (2008).
8. Li, R. *et al. De novo* assembly of human genomes with massively parallel short read sequencing. *Genome Research* **20**, 265–272 (2010).
9. Sims, D., Sudbery, I., Ilott, N. E., Heger, A. & Ponting, C. P. Sequencing depth and coverage: Key considerations in genomic analyses. *Nature Reviews Genetics* **15**, 121–132 (2014).
10. Simpson, J. T. & Pop, M. The theory and practice of genome sequence assembly. *Annual Review of Genomics and Human Genetics* **16**, 153–172 (2015).

11. Phillippy, A. M. New advances in sequence assembly. *Genome Research* **27**, xi–xiii (2017).
12. Bolger, A. M., Lohse, M. & Usadel, B. Trimmomatic: A flexible trimmer for Illumina sequence data. *Bioinformatics* **30**, 2114–2120 (2014).
13. Bankevich, A. *et al.* SPAdes: A new genome assembly algorithm and its applications to single-cell sequencing. *Journal of Computational Biology* **19**, 455–477 (2012).
14. Koren, S. *et al.* Canu: Scalable and accurate long-read assembly via adaptive κ-mer weighting and repeat separation. *Genome Research* **27**, 722–736 (2017).
15. Myers, E. W. *et al.* A whole-genome assembly of *Drosophila*. *Science* **287**, 2196–2204 (2000).
16. Gurevich, A., Saveliev, V., Vyahhi, N. & Tesler, G. QUAST: Quality assessment tool for genome assemblies. *Bioinformatics* **29**, 1072–1075 (2013).
17. Simão, F. A., Waterhouse, R. M., Ioannidis, P., Kriventseva, E. V & Zdobnov, E. M. BUSCO: assessing genome assembly and annotation completeness with single-copy orthologs. *Bioinformatics* **31**, 3210–3212 (2015).
18. Parra, G., Bradnam, K. & Korf, I. CEGMA: A pipeline to accurately annotate core genes in eukaryotic genomes. *Bioinformatics* **23**, 1061–1067 (2007).
19. Hunt, M. *et al.* Circlator: Automated circularization of genome assemblies using long sequencing reads. *Genome Biology* **16**, 294 (2015).

Chapter 7

Exome Sequencing

Setia Pramana,[a] Kwong Qi Bin,[b] Heng Huey Ying,[b]
Nuha Hassim[b] and Ong Ai Ling[b]

[a]Institute of Statistics, Jakarta, Indonesia.
[b]Biotechnology & Breeding Department, Sime Darby Plantation R&D Centre, Selangor, 43400, Malaysia.

Glossary of Terms

SNV: Single Nucleotide Variant. Mutation that occurs on one nucleotide within a genome. The term SNV and SNP are sometimes used interchangeably with the use of SNV targeted at single nucleotide mutation that is less characterized and it is typically rare (e.g. only a single individual is known to have it).

SAM: Sequence Alignment/Map, output format from aligners after mapping of raw reads to a reference genome.

VCF: Variant Call Format, which is the format of a text file that stores SNV.

Introduction

NGS technologies has been in the market since 2004 and they have outperformed the Sanger-based sequencing method. Some applications of NGS include whole genome shotgun sequencing (i.e. WGS) and exome sequencing (i.e. WES), which focus on the entire genome and just the exome portion, respectively. Both WGS and WES generate huge amount of raw data and have similar bioinformatics workflows to extract useful information from them, such as important genetic changes that are associated with human health problems. The technology plays a pivotal role in the new field of personalized medicine, as well as many other important fields of life sciences where DNA sequencing is needed. The focus of this chapter will be on the analysis of whole exome datasets only.

Exome is made up of exons and it represents all expressed genes in a genome. Mutations in exons can lead to changes in the encoded proteins and this can give rise to diseases.[1] In humans, all exons represent about 1% of the genome, but they contain approximately 85% of known disease related variants.[2–4] Given the importance of exons in diseases and other areas, WES and its associated workflow are useful to study.

A major requirement before attempting to perform exome analysis is to have a good quality genome assembled and well annotated genes. Given a reference genome, exome sequencing can be done to identify **SNV** and indel (i.e. insertion and deletion). In order to distinguish exons from other genomic regions, probes are required for this targeted sequencing approach. For this purpose, commercial kits have been designed to specifically capture the exonic sequences.[5] Given the high cost required to assemble a genome of high quality, WES is applied mostly in humans and a few other key crops.[6] In humans, this method has been proven to be effective in medical genetics.[7,8]

An alternative to WES is WGS resequencing of the genome in question. The key advantage of WES over the WGS method is that it generates lesser amount of data, thus making data analysis easier. In addition, the method is useful to deep sequence the target region and it allows for more samples to be sequenced in the same sequencing run. However, with the cost of per base sequence data droppping fast, WGS may soon be the preferred method especially when in the future the cost of acquiring more data is lesser than the cost of the commercial kit needed to capture the exome portion for targeted sequencing. Furthermore, with more data from the WGS method, one is able to capture important genetic variants that are not of exon origin. Moreover, WES is less reliable for the detection of copy number variants (CNVs).[9]

General Workflow of WES

The general workflow for WES is presented in Figure 1. It starts after the sequence reads are produced by a sequencer. The output of the sequencer is usually FASTQ files that contain raw reads of millions of DNA fragments. Filters are then applied to get rid of adaptor sequences, unwanted contaminant sequences and low quality bases. These reads are then aligned to a reference genome, which is provided in FASTA format, using an aligner (e.g. BWA[10], Bowtie2[11] and Novoalign (https://www.novocraft.com/)). The outcome of the alignment is a Sequence Alignment/Map (**SAM**) file and its compressed binary format, a Binary Alignment/Map (BAM) file.

Next, several variant callers such as UnifiedGenotyper or HaplotypeCaller from the Genome Analysis Tool Kit (GATK)[12], SAMtools (mpileup; which is now moved to BCFtools[13]), and Freebayes[14] can be used to find SNPs and indels. As several aligners and variant callers are available, for variant calling in Illumina datasets, aligner BWA-MEM and variant callers SAMtools show best performance.[15] In this chapter, we illustrate the exome-sequencing pipeline using some of the mentioned algorithms.

The mutation obtained are presented in a Variant Call Format (**VCF**) or Binary Variant Call Format (BCF) files, which is then used for

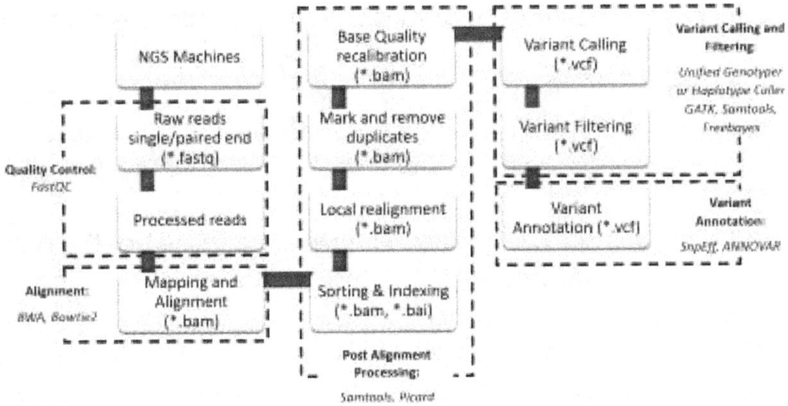

Figure 1. Whole exome sequencing workflow for SNVs detection.

downstream analysis such as annotation of the effects of variants on the encoded proteins using ANNOVAR[16] or SNVEff.[17]

Background Information on the Practical

Dr. James Lupski of the Baylor College of Medicine had his genome sequenced to find out the underlying mutation of Charcot-Marie Tooth (CMT) disease. In the paper Lupski *et al.* 2013, you will find references to all of the raw sequences that were used to analyze his genome.[2] In this practical, we will use Dr. James Lupski's exome data to analyze a NGS workflow that can be used to determine disease-causing mutation. The original raw data has been processed in order to speed up the computation time.

Software

The following software are required to run the analysis of WES:

1. BWA version 0.7.17 (http://bio-bwa.sourceforge.net/). This software maps the raw NGS reads against a large reference genome (e.g. GRCh37, hg19). BWA could map not only short reads (up to 100 bp) but also long reads (up to 1 Mbp). It uses the Burrow-Wheeler Transformation (BWT) algorithm for mapping reads. The input of BWA is a FASTQ file and the output is a BAM file.
2. SAMtools version 1.13 (http://www.htslib.org/). This software provides useful utilities to work with SAM and BAM files. It allows users to view, sort and make index of the BAM/SAM files.
3. BCFtools version 1.13 (http://www.htslib.org/) is 1 of the repositories from SAMtools that is useful for variant calling and manipulating VCF's and BCF's.
4. ANNOVAR (version 2019 Oct24) is a program built for functional annotation of genetic variants acquired from NGS data

and it is written in Perl. To download it, users need to register at http://www.openbioinformatics.org/annovar/annovar_download_form.php.

5. IGV version 2.11.1[18–21] is a visualization tool for SNVs data developed by the Broad Institute, which can be obtained at https://software.broadinstitute.org/software/igv/.

Datasets

- The original exome dataset can be viewed at http://www.ncbi.nlm.nih.gov/sra/?term=SRR866988.
- To speed up this practical, the exome dataset was trimmed from its original 58.8 million paired end reads to just 3714 reads in FASTQ format. These reads were chosen because they mapped around the known causative mutations for the genetic disorder in question here. The processed input data for the practical is input.fq.
- Reference genome: chr5.disease.fasta. This file was extracted from human chromosome 5 at position between 148350000 to 148550000 bp of the genome version GRCh37.

Download Datasets

The datasets can be downloaded at http://bioinfo.perdanauniversity.edu.my/infohub/display/NPB/Index.

Creating a New Folder

```
$ mkdir exome
#All the input files are placed in this folder
$ cd exome
```

Mapping of Raw Data to the Reference Genome

We will be using the BWA program to perform this step.

The first step is to create an index file from the reference genome in order to speed up the mapping process:

```
$ bwa index chr5.disease.fasta
```

This step will produce five files:

```
chr5.disease.fasta.amb
chr5.disease.fasta.ann
chr5.disease.fasta.bwt
chr5.disease.fasta.pac
chr5.disease.fasta.sa
```

Next is the mapping of FASTQ to the reference genome using BWA-MEM (the latest, most recommended for high-quality queries as it is faster and more accurate).

```
$ bwa mem chr5.disease.fasta input.fq > mapped.sam
```

The output would be a mapped.sam file. More information on SAM file is available here:

https://samtools.github.io/hts-specs/SAMv1.pdf

Next, convert the SAM file to a BAM file, this is an essential prerequisite for the following step:

```
$ samtools view -bT chr5.disease.fasta mapped.
sam > mapped.bam
```

It is then followed by sorting and indexing the BAM file:

```
$ samtools sort mapped.bam -o mapped.sort
```

The result is a sorted bam file named mapped.sort

The reference genome needs to be indexed as the beginning step:

```
$ samtools faidx chr5.disease.fasta
```

The result is an index file: chr5.disease.fasta.fai

Variants Calling

BCFtools program will be used for SNVs calling after reference genome mapping. The 'mpileup' command scans and computes all the possible genotypes supported by aligned reads, then calculates the probability of genotypes that are truly present. This is then followed by using the bcftools 'call' command, to identify SNVs and indels, which the output is in VCF as shown in Figure 2.

```
$ bcftools mpileup -f chr5.disease.fasta -O u -o result.bcf mapped.sort
$ bcftools call --multiallelic-caller--variants-only -O v -o result.vcf result.bcf
```

```
#CHROM  POS     ID       REF        ALT     QUAL    FILTER  INFO    FORMAT  mapped.sort
lcl|chr5:148350000-148550000    3464    .       C       A       8.99921 .       DP=1;S(
lcl|chr5:148350000-148550000    9914    .       C       T       6.51248 .       DP=1;S(
lcl|chr5:148350000-148550000    32155   .       C       T       10.7923 .       DP=1;S(
lcl|chr5:148350000-148550000    36526   .       T       G       134.277 .       DP=63;\
lcl|chr5:148350000-148550000    37613   .       A       G       7.30814 .       DP=1;S(
lcl|chr5:148350000-148550000    39764   .       G       A       151.262 .       DP=49;\
lcl|chr5:148350000-148550000    39869   .       T       G       4.96793 .       DP=47;\
lcl|chr5:148350000-148550000    56033   .       C       T       95.4151 .       DP=5;V(
lcl|chr5:148350000-148550000    56387   .       T       C       146.363 .       DP=31;\
lcl|chr5:148350000-148550000    56436   .       G       A       143.353 .       DP=77;\
lcl|chr5:148350000-148550000    57709   .       A       C       136.327 .       DP=12;\
lcl|chr5:148350000-148550000    58102   .       A       G       28.9343 .       DP=18;\
lcl|chr5:148350000-148550000    72282   .       A       G       153.406 .       DP=94;\
lcl|chr5:148350000-148550000    78186   .       G       A       10.7923 .       DP=1;S(
lcl|chr5:148350000-148550000    81854   .       A       G       3.73859 .       DP=2;S(
lcl|chr5:148350000-148550000    86822   .       A       G       10.7923 .       DP=1;S(
lcl|chr5:148350000-148550000    92586   .       T       C       126.395 .       DP=34;\
```

Figure 2. VCF output after SNV calling.

The descriptions of headers in VCF format are as follows:

i. CHROM—chromosome number
ii. POS—position in the genome
iii. ID—SNV identifier
iv. REF—reference allele
v. ALT—alternate allele
vi. QUAL—Phred-scaled quality score for ALT
vii. FILTER—filter status. In this case, we did not set any filter
viii. INFO—additional information

From the result we have detected quite a number of SNVs by mapping the short reads to the reference genome.

These SNVs can further be filtered based on some criteria and thresholds by setting filter options of bcftools view command. For example if we are interested in rare variants, we can select variants with frequency of minor alleles (MAF) < 1%. However, this step is not illustrated here as we have only used a very small FASTQ subset from the original file.

More information about VCF can be found at https://samtools.github.io/hts-specs/VCFv4.2.pdf.

Since we only took 148350000 to 148550000 bp of chromosome 5, for the positions reported from VCF file, we will need to add (148350000-1) bp to it for the actual position. From this example, position 57,709 becomes position 148,407,708 in the original chromosome 5. The command below can calculate the actual position and reformat the VCF file for the next step:

```
$sed 's/lcl|chr5:148350000-148550000/chr5/'
result.vcf > result.vcf1
```

```
$awk '{if ($1 !~ /#/)print ($1 "\t" $2 + 148350000-
1) "\t" substr($0, index($0,$3)); else print $0;}'
result.vcf1 > annovar.input
```

Now we have acquired the SNVs in the right coordinate. The problem now is to find out the effects and functions of these SNVs.

Prediction of SNVs and Indels Effects

For the case of humans, the analysis of SNV effect is simple. The annovar folder should be placed in the exome folder that was created in previous step. To run ANNOVAR, first convert the format from VCF to the required input format using the following command:

```
$ annovar/convert2annovar.pl --format vcf4 --
includeinfo annovar.input > result.annovar
```

Next, to annotate the SNV, use the following Perl Script.

```
$ annovar/annotate_variation.pl --buildver hg19
result.annovar annovar/humandb -outfile SNVs
```

More information on ANNOVAR can be found at http://annovar.openbioinformatics.org/en/latest/user-guide/gene/

Take note that "buildver" is the version of human genome data in use. If the data is of newer version, download and substitute the files in humandb with the newer version from http://hgdownload.cse.ucsc.edu/goldenPath/hg19/database/

The result of the annotation can be found from the output file "SNVs.variant_function" and "SNVs.exonic_variant_function". First, let us take a look at "SNVs.variant_function" file (Figure 3). This file contains the annotation of all variants.

```
intergenic	ADRB2(dist=145266),SH3TC2(dist=8250)	chr5	148353463	148353463	C	A	chr5	1483
intergenic	ADRB2(dist=151716),SH3TC2(dist=1800)	chr5	148359913	148359913	C	T	chr5	1483
UTR3	SH3TC2(NM_024577:c.*2120G>A)	chr5	148382154	148382154	C	T	chr5	148382154	.
exonic	SH3TC2	chr5	148386525	148386525	T	G	chr5	148386525	.	T	G	134.
intronic	SH3TC2	chr5	148387612	148387612	A	G	chr5	148387612	.	A	G
intronic	SH3TC2	chr5	148389763	148389763	G	A	chr5	148389763	.	G	A
exonic	SH3TC2	chr5	148389868	148389868	T	G	chr5	148389868	.	T	G	4.96
intronic	SH3TC2	chr5	148406032	148406032	C	T	chr5	148406032	.	C	T
intronic	SH3TC2	chr5	148406386	148406386	T	C	chr5	148406386	.	T	C
exonic	SH3TC2	chr5	148406435	148406435	G	A	chr5	148406435	.	G	A	143
exonic	SH3TC2	chr5	148407708	148407708	A	C	chr5	148407708	.	A	C	136.
exonic	SH3TC2	chr5	148408101	148408101	A	G	chr5	148408101	.	A	G	28.9
exonic	SH3TC2	chr5	148422281	148422281	A	G	chr5	148422281	.	A	G	153
intronic	SH3TC2	chr5	148428185	148428185	G	A	chr5	148428185	.	G	A
intronic	SH3TC2	chr5	148431853	148431853	A	G	chr5	148431853	.	A	G
```

Figure 3. ANNOVAR output for exonic variant function.

The important fields of this file are the first four columns:

i. First column – representing exonic or intronic SNV
ii. Second column – annotated gene

```
line4   synonymous SNV  SH3TC2:NM_024577:exon16:c.A3594C:p.P1198P,      chr5    148386525       148386525       T
line7   nonsynonymous SNV       SH3TC2:NM_024577:exon14:c.A3292C:p.T1098P,      chr5    148389868       148389868
line10  stopgain        SH3TC2:NM_024577:exon11:c.C2860T:p.R954X,       chr5    148406435       148406435       G
line11  synonymous SNV  SH3TC2:NM_024577:exon11:c.T1587G:p.R529R,       chr5    148407708       148407708       A
line12  synonymous SNV  SH3TC2:NM_024577:exon11:c.T1194C:p.G398G,       chr5    148408101       148408101       A
line13  nonsynonymous SNV       SH3TC2:NM_024577:exon5:c.T505C:p.Y169H, chr5    148422281       148422281       A
line17  nonsynonymous SNV       SH3TC2:NM_024577:exon1:c.A1G:p.M1V,     chr5    148442585       148442585       T
```

Figure 4. ANNOVAR output for variant function.

iii. Third column – chromosome number
iv. Fourth column – base number

As you can see, the SNV list that we have acquired are inclusive of the exonic SNVs (highlighted in red) as reported by Lupski[2], which represents the gene SH3TC2.

The output file "SNVs.exonic_variant_function" contains the variants that are annotated within exonic regions only (Figure 4).

Descriptions of important columns are as follow:

i. First column — line number of this SNV in the original input file (result.annovar).
ii. Second column — the consequence of the variant, possible values are as follow:
    - nonsynonymous SNV — nucleotide change that causes an amino acid change
    - synonymous SNV — nucleotide change that does not cause an amino acid change
    - frame shift — nucleotide insertion/deletion/substitution that causes a frame shift changes in protein coding sequence
iii. Third column — gene symbol: transcript identifier: sequence change in transcript.

The rest of the fields are rather similar to the VCF format. This output file can also be viewed as a Tab-delimited text file in Microsoft Excel.

Visualization

Visualization often provides more information than just text files. To visualize the resulting SNVs acquired, we will be using IGV.

```
# rename the file so that IGV can accept it as
input
$ mv annovar.input result_edited.vcf
$ wget https://data.broadinstitute.org/igv/
projects/downloads/2.11/IGV_Linux_2.11.1_
WithJava.zip
$ unzip IGV_Linux_2.11.1_WithJava.zip
$ sh IGV_Linux_2.11.1/igv.sh
```

Load in result_edited.vcf, and then select chromosome 5. The BAM file that corresponds to the VCF file will be loaded in as well. Make sure that both files are in the same folder.

Key in position chr5:148353463-148542270 in the next tab, as shown in Figure 5.

With this software, we can visualize the location of the SNVs and the genes, together with the mapping quality.

In reality, the analysis of a full exome dataset takes much longer and the main idea of the practical is for users to understand the

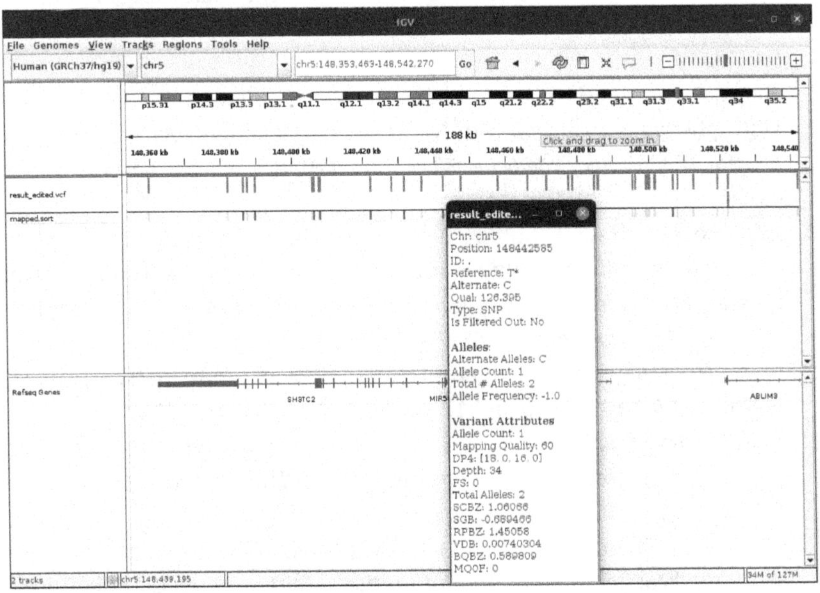

Figure 5. Screenshot of IGV.

basic steps that are required for mining SNVs and indels from an individual.

Conclusion

From the practical, the users have learned how to pre-process exome data starting from FASTQ to getting the VCF file. Prediction of the effects of SNVs and indels is the downstream part of the workflow. One possible application of WES analysis is in the area of personal genomics such as finding the causative mutations in Charcot-Marie Tooth disease.

References

1. Zhang, F. *et al.* Mechanisms for nonrecurrent genomic rearrangements associated with CMT1A or HNPP: rare CNVs as a cause for missing heritability. *American Journal of Human Genetics* **86**, 892–903, doi:10.1016/j.ajhg.2010.05.001 (2010).
2. Lupski, J. R. *et al.* Exome sequencing resolves apparent incidental findings and reveals further complexity of SH3TC2 variant alleles causing Charcot-Marie-Tooth neuropathy. *Genome Medicine* **5**, 57, doi:10.1186/gm461 (2013).
3. Zhu, J. F., Liu, H. H., Zhou, T. & Tian, L. Novel mutation in exon 56 of the dystrophin gene in a child with Duchenne muscular dystrophy. *International Journal of Molecular Medicine* **32**, 1166–1170, doi:10.3892/ijmm.2013.1498 (2013).
4. Zubenko, G. S., Farr, J., Stiffler, J. S., Hughes, H. B. & Kaplan, B. B. Clinically-silent mutation in the putative iron-responsive element in exon 17 of the beta-amyloid precursor protein gene. *Journal of Neuropathology and Experimental Neurology* **51**, 459–463 (1992).
5. Warr, A. *et al.* Exome sequencing: current and future perspectives. *G3 (Bethesda)* **5**, 1543–1550, doi:10.1534/g3.115.018564 (2015).
6. Mascher, M. *et al.* Barley whole exome capture: a tool for genomic research in the genus Hordeum and beyond. *The Plant Journal* **76**, 494–505, doi:10.1111/tpj.12294 (2013).
7. Rabbani, B., Tekin, M. & Mahdieh, N. The promise of whole-exome sequencing in medical genetics. *Journal of Human Genetics* **59**, 5–15, doi:10.1038/jhg.2013.114 (2014).

8. Simons, C. *et al.* Corrigendum: mutations in the voltage-gated potassium channel gene KCNH1 cause Temple-Baraitser syndrome and epilepsy. *Nature Genetics* **47**, 304, doi:10.1038/ng0315-304b (2015).
9. Belkadi, A. *et al.* Whole-genome sequencing is more powerful than whole-exome sequencing for detecting exome variants. *Proceedings of the National Academy of Sciences of the United States of America* **112**, 5473–5478, doi:10.1073/pnas.1418631112 (2015).
10. Li H. and Durbin R. Fast and accurate short read alignment with Burrows-Wheeler Transform. Bioinformatics **25**, 1754–1760, doi:10.1093/bioinformatics/btp324 (2009).
11. Langmead B, Salzberg S. Fast gapped-read alignment with Bowtie 2. Nature Methods **9**, 357–359 (2012).
12. McKenna, A. *et al.* The genome analysis toolkit: a MapReduce framework for analyzing next-generation DNA sequencing data. *Genome Research* **20**, 1297–1303, doi:10.1101/gr.107524.110 (2010).
13. Danecek P, Bonfield JK, Liddle J, Marshall J, Ohan V, Pollard MO, Whitwham A, Keane T, McCarthy SA, Davies RM, Li H, Twelve years of SAMtools and BCFtools. GigaScience, Volume 10, Issue 2, doi:10.1093/gigascience/giab008 (2021).
14. Garrison, E. & Marth, G. Haplotyp e-based variant detection from short-read sequencing. *Genomics* (2012).
15. Hwang, S., Kim, E., Lee, I. & Marcotte, E. M. Systematic comparison of variant calling pipelines using gold standard personal exome variants. *Scientific Reports* **5**, 17875, doi:10.1038/srep17875 (2015).
16. Wang, K., Li, M. & Hakonarson, H. ANNOVAR: functional annotation of genetic variants from high-throughput sequencing data. *Nucleic Acids Research* **38**, e164, doi:10.1093/nar/gkq603 (2010).
17. Cingolani, P. *et al.* A program for annotating and predicting the effects of single nucleotide polymorphisms, SnpEff: SNPs in the genome of Drosophila melanogaster strain w1118; iso-2; iso-3. *Fly* **6**, 80–92, doi:10.4161/fly.19695 (2012).
18. Thorvaldsdottir, H., Robinson, J. T. & Mesirov, J. P. Integrative Genomics Viewer (IGV): high-performance genomics data visualization and exploration. *Briefings in Bioinformatics* **14**, 178–192, doi:10.1093/bib/bbs017 (2013).
19. Robinson, J. T. *et al.* Integrative genomics viewer. *Nature Biotechnology* **29**, 24–26, doi:10.1038/nbt.1754 (2011).
20. James T. Robinson, Helga Thorvaldsdóttir, Aaron M. Wenger, Ahmet Zehir, Jill P. Mesirov. Variant Review with the Integrative Genomics Viewer (IGV).

Cancer Research **77**(21), 31–34, doi:10.1158/0008-5472.CAN-17-0337 (2017).

21. James T. Robinson, Helga Thorvaldsdóttir, Douglass Turner, Jill P. Mesirov. igv.js: an embeddable JavaScript implementation of the Integrative Genomics Viewer (IGV). bioRxiv 2020.05.03.075499, doi:10.1101/2020.05.03.075499 (2020).

Chapter 8

Transcriptomics

Yan Ren,[a] Akzam Saidin[b] and Wai Yee Low[a]

[a]The Davies Research Centre, School of Animal and Veterinary Sciences, University of Adelaide, Roseworthy, SA 5371, Australia
[b]Novocraft Technologies Sdn Bhd, Selangor, Malaysia

Introduction

RNA sequencing (RNA-Seq) is a method that used high-throughput sequencing to decode complementary DNA.[1] Typically, millions of short reads will be produced and these are mapped to a known reference genome for further analysis. The number of reads mapped within exons and/or genes of interest can be quantified and compared across samples.[2,3]

The first step in RNA-Seq analysis is to assess the quality of the reads and to filter out unwanted sequences such as adaptors, primers, and sequences that are not from the target of interest. The raw RNA-Seq data should be inspected with quality control packages such as FASTQC,[3] RSeQC,[4,5] and AfterQC.[6] For a large number of samples, MultiQC[7] may be preferred to compile multiple QC results. Then low-quality reads and unwanted sequences can be removed using tools such as Trimmomatic,[8] Cutadapt,[9] FLEXBAR,[10] and fastp.[11] Low-quality trimming is the removal of low-quality score bases, which usually occur towards the ends of reads. Software such as Trimmomatic comes with commonly used adapters to be screened out, but users can also provide their own sequences for filtering. For a larger dataset, users can consider fastp, which can speed up the cleaning process 2–5 times compared to other tools.[11]

The next step after quality control is either alignment of RNA-seq reads to a reference genome or *de novo* assembly of reads to produce transcripts. RNA-seq alignment will involve mapping the reads to a known reference using aligner programs such as STAR,[12] TopHat2,[13] and HISAT2[14]; HISAT2 is faster than the others. Kallisto[15] is a fast RNA-seq quantification based on pseudo-alignment and requires a transcriptome index. *De novo* assembly of RNA-seq does not rely on having a reference genome or transcriptome but rather it is an attempt to reconstruct larger contiguous sequences by overlapping and merging similar sequences between the reads. Tools such as Trinity[16] and Oases[17] can be used for RNA-seq *de novo* assembly and they have reasonable performances. Recently, SPAdes-rna[18] and BinPacker[19] were developed, which can perform much faster.[20]

RNA-seq is widely used to estimate gene or transcript abundance and to make comparisons across samples obtained under different biological conditions. There are two main strategies for quantifying gene or transcript abundance: "count-based" or "FPKM" (fragments per kilobase of transcript per million mapped reads; paired-end reads)/"RPKM" (reads per kilobase of transcript per million mapped reads; single-end reads). Some researchers reported the TPM (transcripts per million), which is the read per kilobase values divided by the "per million" scaling factor. There is a video from StatQuest that explained this (https://www.rna-seqblog.com/rpkm-fpkm-and-tpm-clearly-explained/). Transcript assembling tools like StringTie[21] and NOIseq[22] report the FPKM/RPKM based abundance value by normalizing the read counts with sequencing depth and gene length. Count based approach estimates abundance by using raw counts from the number of reads that were aligned to the most probable gene (HTSeq[23]).

Differential expression (DE) is a process to identify genes with significant changes in mean expression levels between different conditions such as different tissues, cell types, cells in different environmental conditions or with different genetic backgrounds. There are different models or different approaches in performing DE analysis. The negative binomial model has been adopted in edgeR,[24] DESeq2,[25] limma+voom,[26] and baySeq.[27] However, to better model the discrete distribution of RNA-seq, tools such as SAMseq[28] and NOIseq[29] are using non-parametric models. In terms of different approaches, software like edgeR,[24] DESeq2,[25] and limma+voom[26] measure each genes overall expression per sample but irrespective of the isoforms that may exist.

In the following section, we will go through the basic analysis process of aligning RNA-seq reads to a known reference, measure single sample expression, and perform differential gene expression analysis.

Practical

Datasets & Software

Here, we use Illumina paired-end RNA-Seq data from three different replicates each of Angus and Brahman cattle to study differentially expressed genes between these two breeds. If you have Illumina single-end sequences, please refer to the instructions for running single-end RNA-Seq reads from the manual of the software used here.

Dataset

	Info	URL/GEO accession number
Reference sequence & GTF/GFF3[i]	Brahman chromosome 29	http://ensembl.org/

(*Continued*)

[i]Genome assembly of Brahman (Bos indicus) and its GTF/GFF3 file were obtained from ENSEMBL (https://asia.ensembl.org/Bos_indicus_hybrid/Info/Index).

(*Continued*)

	Info	URL/GEO accession number
Read set[ii]	Angus_RNA_seq_1(liver)	GSM4485877 (Liver-sample-7)
	Angus_RNA_seq_2(liver)	GSM4485870 (Liver-sample-60)
	Angus_RNA_seq_3(liver)	GSM4485868 (Liver-sample-53)
	Brahman_RNA_seq_1(liver)	GSM4485867 (Liver-sample-22)
	Brahman_RNA_seq_2(liver)	GSM4485881 (Liver-sample-65)
	Brahman_RNA_seq_3(liver)	GSM4485872 (Liver-sample-99)
Adaptor sequences[iii]	Illumina adaptor sequences	AGATCGGAAGAGC

Software Required

Software	Version	URL
Data download		
Sratoolkit	2.11.0	https://github.com/ncbi/sra-tools/wiki
QC and Reads Preprocessing		

[ii] Illumina HiSeq 4000 double-ended RNA-Seq reads for cattle liver tissues were obtained from NCBI (https://www.ncbi.nlm.nih.gov/geo/query/acc.cgi?acc=GSE148909).

[iii] Adaptor sequences were obtained from Illumina support page (https://support-docs.illumina.com/SHARE/AdapterSeq/Content/SHARE/AdapterSeq/AdapterSequencesIntro.htm).

(*Continued*)

Software	Version	URL
FastQC[3]	0.11.9	http://www.bioinformatics.babraham.ac.uk/projects/fastqc/
multiqc[7]	1.11	https://multiqc.info
Cutadapt[9]	3.4	https://cutadapt.readthedocs.io/en/stable/
Trim Galore (a wrapper around Cutadapt)	0.6.5	https://github.com/FelixKrueger/TrimGalore
Reads Alignment		
HISAT2[14]	2.2.1	http://daehwankimlab.github.io/hisat2/
SAMTOOLS[17]	1.13	https://github.com/samtools/samtools
IGV	2.3	https://software.broadinstitute.org/software/igv/
Expression Analysis		
featureCounts[30]	1.22.2	https://www.rdocumentation.org/packages/Rsubread/versions/1.22.2/topics/featureCounts
HTSeq[23]	0.13.5	https://pypi.python.org/pypi/HTSeq
pysam[31]	0.16.0.1	https://github.com/pysam-developers/pysam
R[32]	3.2.3	https://www.r-project.org/
edgeR[24]	3.8.6	https://bioconductor.org/packages/release/bioc/html/edgeR.html
limma[33]	3.13	https://bioconductor.org/packages/release/bioc/html/limma.html

Note: Installation instructions for each software can be found in the respective download URLs.

Reads Pre-processing & Quality Control (QC)

Prepare Files

Sequence Read Archive[34] (SRA) is the largest publicly available repository of high throughput sequencing data, which is available through multiple cloud providers and NCBI servers. To download the data from the SRA, we installed and configured (https://github.com/ncbi/sra-tools/wiki/03.-Quick-Toolkit-Configuration) the sratoolkit.

```
# current working directory is called MAINDIR and recorded the path.
MAINDIR=${PWD}
# can run following code to back to the MAINDIR anytime if needed
cd ${MAINDIR}
```

We create a folder SRA_data that sets as the "Location of user-repository" in the configuration.

```
mkdir SRA_data
```

In the following commands, we create another directory (liver_raw_RNA_seq), prefetch the SRA files into the SRA_data/sra, and then dump the raw FASTQ files into the liver_raw_RNA_seq folder. For paired-end reads, the "split-files" argument is needed in the fastq-dump command.

```
mkdir liver_raw_RNA_seq

mkdir liver_raw_RNA_seq/Angus_RNA_seq_1
for n in {76..80}; do prefetch -v SRR115695${n}; done
for n in {76..80}; do fastq-dump —outdir liver_raw_RNA_seq/Angus_RNA_seq_1 \
--split-files SRA_data/sra/SRR115695${n}.sra; done

mkdir liver_raw_RNA_seq/Angus_RNA_seq_2
for n in {42..46}; do prefetch -v SRR115695${n}; done
for n in {42..46}; do fastq-dump —outdir liver_raw_RNA_seq/Angus_RNA_seq_2 \
--split-files SRA_data/sra/SRR115695${n}.sra; done

mkdir liver_raw_RNA_seq/Angus_RNA_seq_3
for n in {32..36}; do prefetch -v SRR115695${n}; done
for n in {32..36}; do fastq-dump —outdir liver_raw_RNA_seq/Angus_RNA_seq_3 \
--split-files SRA_data/sra/SRR115695${n}.sra; done

mkdir liver_raw_RNA_seq/Brahman_RNA_seq_1
for n in {27..31}; do prefetch -v SRR115695${n}; done
for n in {27..31}; do fastq-dump —outdir liver_raw_RNA_seq/Brahman_RNA_seq_1 \
--split-files SRA_data/sra/SRR115695${n}.sra; done
```

```
mkdir liver_raw_RNA_seq/Brahman_RNA_seq_2
for n in {88..92}; do prefetch -v SRR115695${n}; done
for n in {88..92}; do fastq-dump -outdir liver_raw_RNA_seq/Brahman_RNA_seq_2 \
--split-files SRA_data/sra/SRR115695${n}.sra; done

mkdir liver_raw_RNA_seq/Brahman_RNA_seq_3
for n in {52..56}; do prefetch -v SRR115695${n}; done
for n in {52..56}; do fastq-dump -outdir liver_raw_RNA_seq/Brahman_RNA_seq_3 \
--split-files SRA_data/sra/SRR115695${n}.sra; done
```

To make the practical easier to follow, we have renamed the files accordingly:

```
cd liver_raw_RNA_seq

for f in *; do
    if [ -d "$f" ]; then
        cat ${f}/*_R1*> ./${f}_R1.fastq
        cat ${f}/*_R2*> ./${f}_R2.fastq
    fi
done
```

Notice that we have three biological replicates for Angus and Brahman.

Perform Initial QC

We make a folder and produce all FastQC reports into the folder using 4 CPU cores.

```
mkdir raw_FastQC_Report
fastqc -o raw_FastQC_Report -t 4 *.fasta
```

FastQC will produce a html file for each FASTA file. These files can be opened using a web browser. Using multiqc, multiple FastQC reports can be summarised into a single html file.

```
multiqc ./raw_FastQC_Report

# to back to the main directory
cd ${MAINDIR}
```

Trimming for Bad Quality and Adapters

Based on the initial quality reports, trim_galore can be used to trim the bases (e.g. those with <Q20 or <Q10; Q is the Phred quality score). If there are adaptors in your read set, you can trim the adaptor sequences using read trimming tools like cutadapt,[9]

trimmomatic,[8] AdapterRemoval,[35] and flexbar.[36] The adaptor sequences for your RNA-seq library can be obtained from your sequencing provider. In this example, we will use trim_galore (a wrapper around Cutadapt) to remove low-quality segments of the reads and adaptor sequence(s) (if any).

```
mkdir trim_galore_results
for n in {Angus_RNA_seq_1 Angus_RNA_seq_2 Angus_RNA_seq_3 Brahman_RNA_seq_1 Brahman_RNA_seq_2
Brahman_RNA_seq_3}
do

 trim_galore --quality 10 --length 70 -o trim_galore_results --paired raw_FastQC_
Report/${n}_1.fastq raw_FastQC_Report/${n}_2.fastq
done
```

The trimming report will be generated for each file and it can be viewed in command line with less. As an example, we view the trimming report for file Angus_RNA_seq_1_R2.fastq.

```
cd trim_galore_results
less Angus_RNA_seq_1_R2.fastq_trimming_report.txt
```

The report is quite informative as shown below.

```
=== Summary ===

Total reads processed:              49,189,375
Reads with adapters:                24,346,613 (49.5%)
Reads written (passing filters):    49,189,375 (100.0%)

Total basepairs processed:  4,918,937,500 bp
Quality-trimmed:               14,421,592 bp (0.3%)
Total written (filtered):   4,709,092,207 bp (95.7%)

=== Adapter 1 ===

Sequence: AGATCGGAAGAGC; Type: regular 3'; Length: 13; Trimmed: 24346613 times.

No. of allowed errors:
0-9 bp: 0; 10-13 bp: 1

Bases preceding removed adapters:
  A: 23.4%
  C: 32.6%
  G: 30.6%
  T: 13.4%
  none/other: 0.0%

Overview of removed sequences
length     count      expect     max.err    error counts
1          9373549    12297343.8 0          9373549
2          3352314    3074335.9  0          3352314
3          1411994    768584.0   0          1411994
…

Total number of sequences analysed for the sequence pair length validation: 49189375

Number of sequence pairs removed because at least one read was shorter than the length cutoff
(70 bp): 1437289 (2.92%)
```

From the trimming report file, it is reported that 49.5% of the read pairs contain adapters, and 95.7% of the bases passed the quality filter. If no adapter sequence is given in the command (use–adapter), the trim_galore can automatically detect the sequencing provider and use the default adapters. The adapters used and the overview of removed sequences are recorded in the report.

Run FastQC on Trimmed Reads

You can skip this step if you run trim_galore with flag –fastqc, as it will automatically run FastQC once trimming is complete. Another option is to run the following code for FastQC to check sequence quality again after trimming.

```
mkdir trimed_FastQC_Report
fastqc -o trimed_FastQC_Report -t 4 *val_*.fq
multiqc ./trimed_FastQC_Report
# to back to the main directory
cd ${MAINDIR}
```

Compare the FastQC reports before and after trimming of reads and note the improvement.

HISAT2: Reads Alignment

For this practical, we will use a graph-based data aligner, HISAT2, for reads alignment.

Prepare Files

Create a directory for your reference files.

```
mkdir UOA_Brahman_1 _chr29
```

In order to make the process running faster, we only use the reference for chromosome 29 in this practice. In a real analysis, all chromosomes should be used to allow reads to be mapped to them. Download cattle chromosome 29 reference sequence file (fasta) and gene feature file (gff/gtf) from the ENSEMBL ftp site. You will need to generate chromosome 29 gtf file from the main

Angus gfff/gtf file. You can extract the features with the following command:

```
zgrep -P '^29\t' Bos_indicus_hybrid.UOA_Brahman_1.104.gtf.gz > UOA_Brahman_1 _chr29/Brahman_
chr29.gtf

# alternatively, can use following code:
unzip Bos_indicus_hybrid.UOA_Brahman_1.104.gtf.gz
grep -P '^29\t' Bos_indicus_hybrid.UOA_Brahman_1.104.gtf > UOA_Brahman_1 _chr29/Brahman_
chr29.gtf
```

Generate Genome Index

Before performing RNA reads alignment, the reference sequence needs to be indexed.

```
hisat2-build Bos_indicus_hybrid.UOA_Brahman_1.cds.all.fa.gz UOA_Brahman_1 _chr29/ Brahman_
chr29.ref
```

Reads Alignment

Next, we will perform reads alignment using HISAT2 aligner and generate the respective alignment file for each biological replicates of Brahman and Angus cattle.

```
mkdir HISAT2_alignment_results

for n in { Angus_RNA_seq_1 Angus_RNA_seq_2 Angus_RNA_seq_3 Brahman_RNA_seq_1 Brahman_RNA_
seq_2 Brahman_RNA_seq_3}
do

       hisat2 -q --threads 4 -x UOA_Brahman_1 _chr29/Brahman_chr29 -1 trim_galore_
results/${n}_1_val_1.fq -2 trim_galore_results/${n}_2_val_2.fq -S ${n}.sam
       samtools view -bS ${n}.sam > ${n}.bam
       samtools sort ${n}.sam -o ${n}.sorted.bam
       samtools index ${n}.sorted.bam
       rm ${n}.sam ${n}.bam
done

mv *sorted.bam HISAT2_alignment_results
```

You can view reads aligned to the expressed gene regions by loading the BAM files into IGV. Expressed gene regions are shown on the first line and the reads are in grey and green (Figure 1).

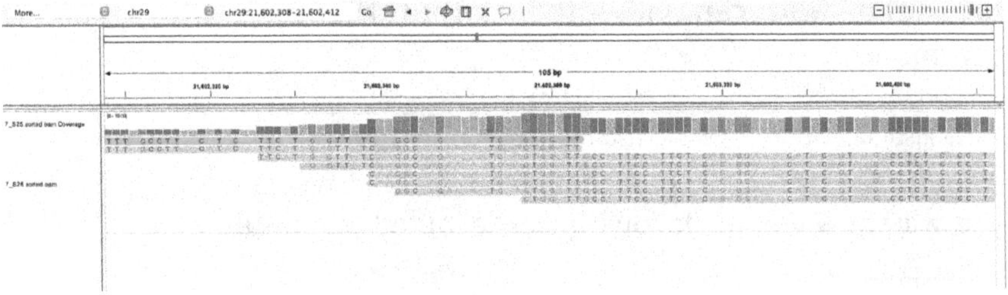

Figure 1. Visualization of reads aligned to the expressed region using IGV.

Single Sample Expression

To obtain individual gene expression counts, each of the BAM files were processed with a gene expression tool. As mentioned before, there are two approaches in quantifying gene expression, which are either "raw counts" or FPKM. In this practical, we will go through the basic for both methods using HTSEQ, featureCounts and StringTie.

Gene Expression Count Using HTSEQ

Following command can be used to perform "raw counts" based gene expression analysis using HTSEQ.

```
# pysam need to be installed in the same environment as HTSEQ for reading bam files

mkdir HTSEQ_results

for n in {Angus_RNA_seq_1 Angus_RNA_seq_2 Angus_RNA_seq_3 Brahman_RNA_seq_1 Brahman_RNA_seq_2
Brahman_RNA_seq_3}
do
    htseq-count --format bam --order pos --mode intersection-strict --stranded reverse
--minaqual 1 --type exon --idattr gene_id ${n}.sorted.bam UOA_Brahman_1_chr29/Brahman_chr29.
gtf > HTSEQ_results/{n}.sorted.gene_read_counts_table.tsv
done
```

In the comment, flag –format is for bam file, --order is to order the results by positions, --mode intersection-strict is to separate all the sets when dealing with overlapping features.

Htseq-count generates a 2 columns file, the first column is for gene id and the second column is for the count of reads mapped to the gene. You can view the output file with the following command:

```
head HTSEQ_results/Angus_RNA_seq_1.sorted.gene_read_counts_table.tsv
```

This will show you the following tab delimited text result:

```
ENSBIXG00005000008    1267
ENSBIXG00005000048    26
ENSBIXG00005000119    0
ENSBIXG00005000121    0
ENSBIXG00005000187    30
ENSBIXG00005000194    204
ENSBIXG00005000218    0
ENSBIXG00005000222    5
ENSBIXG00005000245    192
ENSBIXG00005000306    19
```

The last 5 lines of the tsv file generated by htseq-count are basic statistics of feature type. We can view them by using the tail command.

```
tail HTSEQ_results/Angus_RNA_seq_1.sorted.gene_read_counts_table.tsv
```

The basic statistics by htseq-count are marked with __ in front of the feature description.

```
ENSBIXG00005029125          1
ENSBIXG00005029130          30
ENSBIXG00005029163          629
ENSBIXG00005029237          0
ENSBIXG00005031433          0
__no_feature                50115931
__ambiguous                 31050
__too_low_aQual             1073658
__not_aligned               1506744
__alignment_not_unique      4195983
```

Gene Expression Count Using featureCounts

featureCounts is another software that produces "raw counts", and it can be run as either a command in Linux- or R-based environments. Here is an example to make a list of bam files and run featureCounts through all the files use 4 cores:

```
mkdir featureCounts_results
sampleList=`find HISAT2_alignment_results -name "*.sorted.bam" | tr '\n' ' '`

featureCounts -T 4 -a UOA_Brahman_1_chr29/ Brahman_chr29.gtf -o featureCounts_results/counts.
out ${sampleList}
```

Rstudio can be used to load the R environment. We can use the following code to perform featureCounts in R:

```
library(Rsubread)

sortbam <- dir(pattern=".sorted.bam")

countsensembl <- featureCounts(sortbam, annot.ext=" UOA_Brahman_1_chr29/Brahman_chr29.gtf
", isGTFAnnotationFile=T, GTF.featureType="gene", GTF.attrType="gene_id", isPairedEnd=TRUE,
reportReads=NULL)
```

Within R/Rstudio, we can use the following code to see the target bam files, gene counts, gene annotation and statistics.

```
countsensembl$target
countsensembl$counts
countsensembl$annotation
countsensembl$stat
```

Gene Expression Count StringTie

To perform FPKM-based gene expression, we can run StringTie with the following command:

```
# make sure the same naming convention between GTF and the genome sequences
sed "s/^29/CM011832.1/g" -i UOA_Brahman_1_chr29/ Brahman_chr29.gtf

mkdir StringTie_results

for n in {Angus_RNA_seq_1 Angus_RNA_seq_2 Angus_RNA_seq_3 Brahman_RNA_seq_1 Brahman_RNA_seq_2
Brahman_RNA_seq_3}
          do

     mkdir StringTie_results/${n}
     stringtie ${n}.sorted.bam -e -B -p 4 -G UOA_Brahman_1_chr29/ Brahman_chr29.gtf -o
StringTie_results/${n}/${n}.stdout
done
```

Individual gene expression counts were calculated by StringTie and reported in FPKM format. Each of the fields reported are as explained below in t_data.ctab:

Column name	Description
t_id	1 to the number of identical transcripts in the GFF/GTF file
chr	Chromosome of where the reference transcript located
strand	Reading direction of the reference transcript
start	Start site of the reference transcript
end	end site of the reference transcript
t_name	A unique identifier describing the object (gene, transcript, CDS, primary transcript)
num_exons	Number of exons of the reference transcript
length	The number of base pairs in the transcript, or '-' if not a transcript/primary transcript
gene_id	The gene_id(s) associated with the object
gene_name	The gene_short_name(s) associated with the object
cov	Estimate for the absolute depth of read coverage across the object
FPKM	FPKM of the object in sample 0

Differential Expression

For DE analysis, we prefer to start with the "raw counts" method. limma and edgeR are two powerful packages to perform DEG analysis on "raw counts". In most cases, FPKM can be converted back to "raw counts" using the length of genes, and then one can perform DEG analysis without voom, edgeR, and DESeq (detailed in the limma vignette). If FPKM is all you have, you can $\log_2$ convert the values and the perform limma differential analysis using eBayes() with trend=TRUE. Normally, you may want to add a 0.1 (i.e. log2(FPKM+0.1)) to avoid applying log to zero values. Here we focus on the DEG analysis for counts data.

edgeR & Limma for Counts Data

Make a counts table

For HTSEQ, we will need to join the tsv files generated by htseq-count to create a single main table. To merge multiple files, first we make a temporary file and write the first file to it. Then, we run a for loop to merge the other files sequentially.

```
cat Angus_RNA_seq_1.sorted.gene_read_counts_table.tsv > gene_counts_HTseq.gff

for n in {Angus_RNA_seq_2 Angus_RNA_seq_3 Brahman_RNA_seq_1 Brahman_RNA_seq_2 Brahman_RNA_
seq_3}
do
     join gene_counts_HTseq.gff ${n}.sorted.gene_read_counts_table.tsv > temp.gff
     cat temp.gff > gene_counts_HTseq.gff
done

rm temp.gff

tail gene_counts_HTseq.gff
```

gene_counts_HTseq.gff has 5 extra lines which are the basic statistics. These lines must be removed prior to using it as an input file for edgeR. This can be removed using the following command:

```
sed '/^_/d' gene_counts_HTseq.gff > gene_counts_HTseq.tab
```

Similarly, in command line, one can run the following code to produce a single count table from the featureCounts results.

```
cut -f1,7- counts.out | sed 1d > genes.out
```

Now the file is ready to be used by edgeR to make a DGEList formatted object, and then apply limma linear model for comparison between groups.

Perform edgeR and limma and linear model

Select the right code for the software that you used to read the count table into R environment:

```
# for HTSEQ
gene_counts <- read.table("gene_counts_HTseq.tab", row.names=1, quote="\"", comment.char="")

# for featureCounts (command line)
gene_counts <- read.table("featureCounts_results/genes.out",header = T, row.names=1,
quote="\"", comment.char="")

# for featureCounts (R version)
gene_counts <- countsensembl$counts
```

To set the right column names:

```
colnames(gene_counts) <- c("Angus_RNA_seq_1", "Angus_RNA_seq_2", "Angus_RNA_seq_3", "Brahman_
RNA_seq_1", "Brahman_RNA_seq_2", "Brahman_RNA_seq_3")
```

After having the count table, we need to read the annotation file.

```
library(rtracklayer)

Brahman_chr29_anno <- rtracklayer::import('Brahman_chr29.gtf')
Brahman_chr29_anno <- as.data.frame(Brahman_chr29_anno)
```

Now we are ready to run edgeR and limma linear model. Here is a simple model to fit a just one explanatory factor variable, i.e. the cattle subspecies (Angus or Brahman). As a reminder, the expression data was from liver and female cattle. More complex model can be set and please refer to the limma manual.

```
library(edgeR)
library(limma)

# make master DGEList

DEG <- list()

# set the count
DEG$counts <- apply(gene_counts,MARGIN = 2,FUN = as.numeric) %>%
  set_rownames(rownames(gene_counts))

DEG$anno <- subset(Brahman_chr29_anno, gene_id %in% rownames(DEG$counts))
DEG$anno <- subset(Brahman_chr29_anno, type %in% "gene")

# set expression level cutoff, can change from 0.5 to 1 for RNAseq data.
Sel <- rowSums(cpm(DEG$counts) > 0.5) >= 3
DEG$counts <- DEG$counts[sel,]
DEG$anno <- DEG$anno[sel,]
DEG <- new("DGEList", DEG)
```

We used the trimmed mean of M-values (TMM) method[37] in edgeR as the normalization method.

```
# normalization
DEG <- calcNormFactors(DEG, method="TMM")
```

Set a matrix to specific the Angus and Brahman samples, and build a design matrix for regression model with the specified formula and data.

```
# set design matrix
phenotype <- as.factor(rep(c("Angus", "Brahman"),each= 3))
modelpheno <- model.matrix(~0+phenotype)

#combine the DEGList and the model

despheno <- estimateDisp(DEG,modelpheno)

# show the dispersion
sqrt(despheno$common.dispersion)
```

By using sqrt(), we can see the dispersion is 0.2149514, then we can plot the biological coefficient variation with the following code:

```
plotBCV(despheno)
```

In a real analysis, all chromosomes should be used instead of just chromosome 29 (Figure 2) and we show the plot in Figure 3.

Each dot on the plot represents one gene. Theoretically, we expect to see all the dots located close to the blue line, and the trend line tail ends up parallel to the red line.

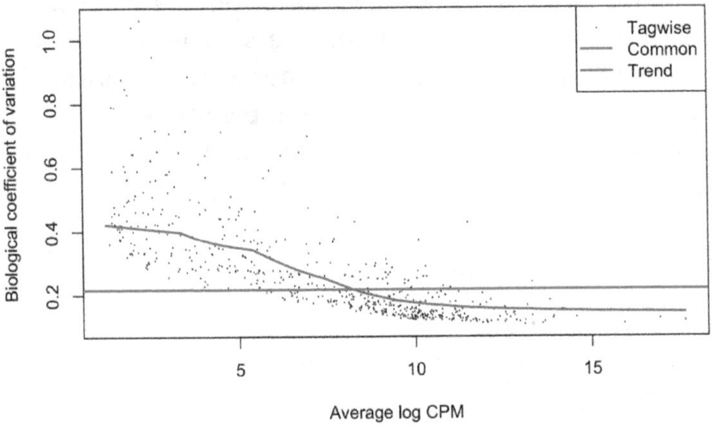

Figure 2. Biological coefficient of the variations for Brahman chromosome 29.

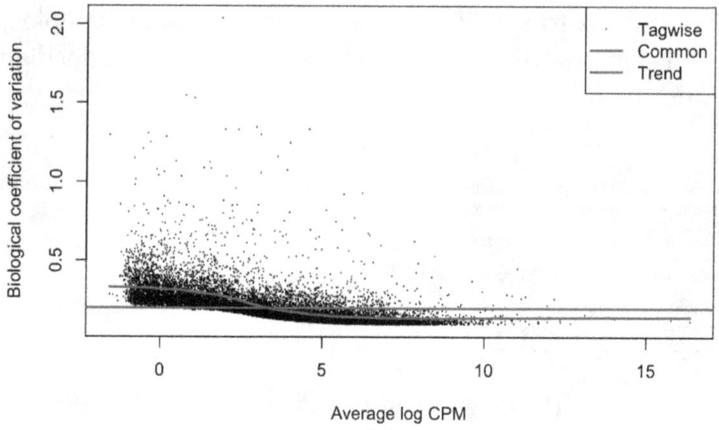

Figure 3. Biological coefficient of the variations across all Brahman chromosomes.

```
#use voom observational-level weights to quantity sample quality

vpheno <- voomWithQualityWeights(despheno,design=modelpheno,normalize.method = "none",
plot=T, col=as.numeric(phenotype))

#give the colour labels for Angus and Brahman
cols <- rep("red", 6)
cols[phenotype=="Angus"] <- "black"

#mds plot
plotMDS(vpheno, label=phenotype, col=cols, dim.plot=c(1,2), main="RNA-seq MDSplot")
```

In the MDS plot, biological replicates from the same breed clustered together but not so well for one sample from each of the breeds (Figure 4). When all chromosomes were used instead of just chromosome 29, only one Brahman sample seemed not to cluster well with the other two Brahman samples in dimension 2 (Figure 5).

```
colnames(modelpheno)=c("Angus", "Brahman")

# make a contrast
contrpheno <- makeContrasts("Angus-Brahman"=Angus-Brahman, levels=modelpheno)

# fit linear model
lmFitpheno <- lmFit(vpheno)

colnames(lmFitpheno$coefficients) <- rownames(contrpheno)

# fit a contrast
vfitcpheno <- contrasts.fit(lmFitpheno,contrpheno)

# apply the Bayes moderation
vfitcpheno <- eBayes(vfitcpheno)
```

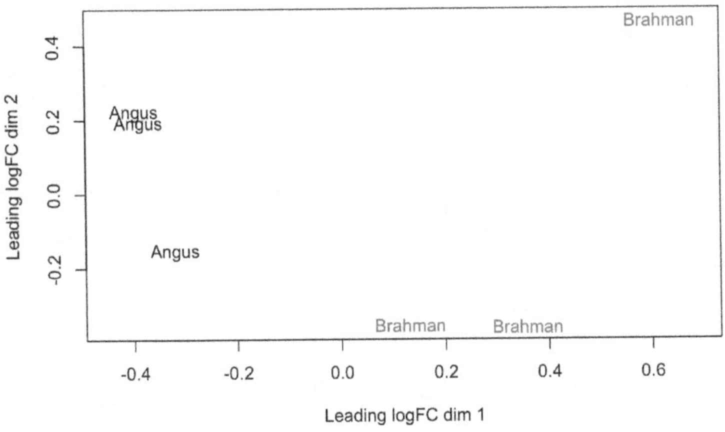

Figure 4. MDS plot for Brahman chromosome 29.

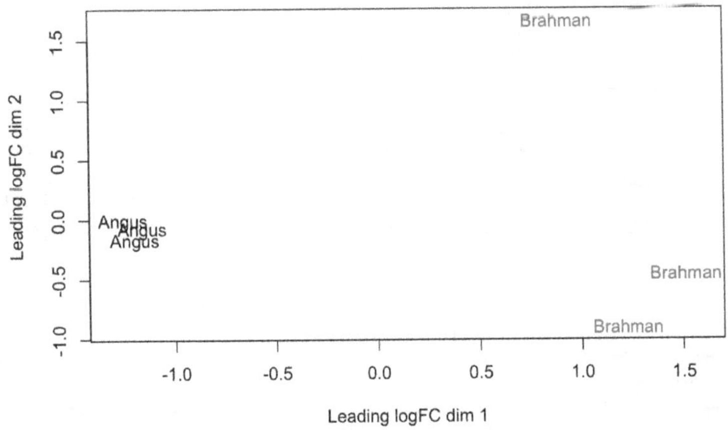

Figure 5. MDS plot for all Brahman chromosomes.

By showing the summary table and the results of differential analysis, we can see there are two genes significantly down-regulate on chromosome 29 (*p.value=0.05, logFC = 1*). In a real case with all chromosomes, there are more differentially expressed genes, and the number can be adjusted by setting different cut off for p.value and lfc.

```
summary(decideTests(vfitcpheno,p.value=0.05, lfc = 1))

Angus-Brahman
Down       2
NotSig   574
Up         0
```

The function topTable()shows the information of the two differentially expressed genes, and here we use False discovery rate to adjust the p-values.

```
topTable(vfitcpheno,n=Inf,coef=1,p.value=0.05,lfc = 1, method="fdr")
```

Plots and tables that were produced in the process can be viewed using Rstudio viewer/table panels. By running the View() function, we can see the following table. See Table 1 for an explanation of the column names.

gene_id	chr	gene_ name	gene_ biotype	logFC	AveExpr	*t*	*P.vaule*	adj.*P*.Val	*B*
ENSBIXG00005012881	29	*NA*	snoRNA	6.844771	4.987784	8.497328	1.610486e-05	0.009276397	3.161074
ENSBIXG00005012872	29	*NA*	snoRNA	4.814734	4.366816	7.756628	3.288804e-05	0.009471755	2.635556

The method used *t*-statistics to decide the *p*-values and it adjusted the *p*-values for multiple testing by setting method="fdr" in the topTable()function.

Table 1. Differentially expressed gene results.

Column Name	Description
gene_id	ENSEMBL gene ID from the GFF/GTF file
chr	Chromosome of where the reference transcript located
gene_name	ENSEMBL gene names from the GFF/GTF file
gene_biotype logFC	ENSEMBL gene type from the GFF/GTF file
AveExpr	The values of log2 fold changes.
	Average expression values
t	Values of *t*-statistics to decide the significance of DE
	P values
P.value	Adjusted *P* values (*Q* values)
adj.*P*.Val	Values of *B*-statistic to decide whether DE has occurred[38]
B	

The smaller the adjusted *p*-values the more significant the difference.

References

1. Kukurba, K. R. & Montgomery, S. B. RNA sequencing and analysis. *Cold Spring Harbor Protocols* **11**, 951–969 (2015).
2. Soneson, C. & Delorenzi, M. A comparison of methods for differential expression analysis of RNA-seq data. *BMC Bioinformatics* **14**, 91 (2013).
3. Oshlack, A., Robinson, M. D. & Young, M. D. From RNA-seq reads to differential expression results. *Genome Biology* **11**, 220 (2010).
4. Wang, L., Wang, S. & Li, W. RSeQC: quality control of RNA-seq experiments. *Bioinformatics* **28**, 2184–2185 (2012).
5. Wang, L. *et al.* Measure transcript integrity using RNA-seq data. *BMC Bioinformatics* **17**, 58 (2016).
6. Chen, S. *et al.* AfterQC: automatic filtering, trimming, error removing and quality control for fastq data. *BMC Bioinformatics* **18**, 80 (2017).
7. Ewels, P., Magnusson, M., Lundin, S. & Käller, M. MultiQC: summarize analysis results for multiple tools and samples in a single report. *Bioinformatics* **32**, 3047–3048 (2016).
8. Bolger, A. M., Lohse, M. & Usadel, B. Trimmomatic: a flexible trimmer for Illumina sequence data. *Bioinformatics* **30**, 2114–2120 (2014).
9. Martin, M. Cutadapt removes adapter sequences from high-throughput sequencing reads. *EMBnet. journal* **17**, 10 (2011).
10. Dodt, M., Roehr, J. T., Ahmed, R. & Dieterich, C. FLEXBAR-flexible barcode and adapter processing for next-generation sequencing platforms. *Biology (Basel)* **1**, 895–905 (2012).
11. Chen, S., Zhou, Y., Chen, Y. & Gu, J. fastp: an ultra-fast all-in-one FASTQ preprocessor. *Bioinformatics* **34**, i884–i890 (2018).
12. Dobin, A. *et al.* STAR: ultrafast universal RNA-seq aligner. *Bioinformatics* **29**, 15–21 (2013).
13. Kim, D. *et al.* TopHat2: accurate alignment of transcriptomes in the presence of insertions, deletions and gene fusions. *Genome Biology* **14**, R36 (2013).
14. Kim, D., Paggi, J. M., Park, C., Bennett, C. & Salzberg, S. L. Graph-based genome alignment and genotyping with HISAT2 and HISAT-genotype. *Nature Biotechnology* **37**, 907–915 (2019).
15. Bray, N. L., Pimentel, H., Melsted, P. & Pachter, L. Near-optimal probabilistic RNA-seq quantification. *Nature Biotechnology* **34**, 525–527 (2016).

16. Haas, B. J. *et al.* De novo transcript sequence reconstruction from RNA-seq using the Trinity platform for reference generation and analysis. *Nature Protocols* **8**, 1494–1512 (2013).
17. Schulz, M. H., Zerbino, D. R., Vingron, M. & Birney, E. Oases: robust de novo RNA-seq assembly across the dynamic range of expression levels. *Bioinformatics* **28**, 1086–1092 (2012).
18. Bankevich, A. *et al.* SPAdes: a new genome assembly algorithm and its applications to single-cell sequencing. *Journal of Computational Biology* **19**, 455–477 (2012).
19. Liu, J. *et al.* BinPacker: packing-based de novo transcriptome assembly from RNA-seq data. *PLoS Computational Biology* **12**, e1004772 (2016).
20. Hölzer, M. & Marz, M. De novo transcriptome assembly: A comprehensive cross-species comparison of short-read RNA-Seq assemblers. *Gigascience* **8**, (2019).
21. Pertea, M. *et al.* StringTie enables improved reconstruction of a transcriptome from RNA-seq reads. *Nature Biotechnology* **33**, 290–295 (2015).
22. Tarazona, S., García, F., Ferrer, A., Dopazo, J. & Conesa, A. NOIseq: a RNA-seq differential expression method robust for sequencing depth biases. *EMBnet. journal* **17**, 18 (2012).
23. Anders, S., Pyl, P. T. & Huber, W. HTSeq — a Python framework to work with high-throughput sequencing data. *Bioinformatics* **31**, 166–169 (2015).
24. Robinson, M. D., McCarthy, D. J. & Smyth, G. K. edgeR: a Bioconductor package for differential expression analysis of digital gene expression data. *Bioinformatics* **26**, 139–140 (2010).
25. Love, M. I., Huber, W. & Anders, S. Moderated estimation of fold change and dispersion for RNA-seq data with DESeq2. *Genome Biology* **15**, 550 (2014).
26. Ritchie, M. E. *et al.* limma powers differential expression analyses for RNA-sequencing and microarray studies. *Nucleic Acids Research* **43**, e47 (2015).
27. Hardcastle, T. J. & Kelly, K. A. baySeq: empirical Bayesian methods for identifying differential expression in sequence count data. *BMC Bioinformatics* **11**, 422 (2010).
28. Li, J. & Tibshirani, R. Finding consistent patterns: a nonparametric approach for identifying differential expression in RNA-Seq data. *Statistical Methods in Medical Research* **22**, 519–536 (2013).
29. Tarazona, S. *et al.* Data quality aware analysis of differential expression in RNA-seq with NOISeq R/Bioc package. *Nucleic Acids Research* **43**, e140 (2015).

30. Liao, Y., Smyth, G. K. & Shi, W. featureCounts: an efficient general purpose program for assigning sequence reads to genomic features. *Bioinformatics* **30**, 923–930 (2014).
31. Li, H. *et al.* The sequence alignment/map format and SAMtools. *Bioinformatics* **25**, 2078–2079 (2009).
32. Ihaka, R. & Gentleman, R. R: a language for data analysis and graphics. *Journal of Computational and Graphical Statistics* **5**, 299 (1996).
33. Smyth, G. K. In: R. Gentleman, V. J. Carey, W. Huber, R. A. Irizarry, & S. Dudoit (eds.), *Bioinformatics and Computational Biology Solutions Using R and Bioconductor*, 397–420. Springer (2005). doi:10.1007/0-387-29362-0_23.
34. Leinonen, R., Sugawara, H., Shumway, M. & International Nucleotide Sequence Database Collaboration. The sequence read archive. *Nucleic Acids Research* **39**, D19–D21 (2011).
35. Schubert, M., Lindgreen, S. & Orlando, L. AdapterRemoval v2: rapid adapter trimming, identification, and read merging. *BMC Research Notes* **9**, 88 (2016).
36. Dodt, M., Roehr, J. T., Ahmed, R. & Dieterich, C. FLEXBAR — flexible barcode and adapter processing for next-generation sequencing platforms. *Biology (Basel)* **1**, 895–905 (2012).
37. Robinson, M. D. & Oshlack, A. A scaling normalization method for differential expression analysis of RNA-seq data. *Genome Biology* **11**, R25 (2010).
38. Lönnstedt, I. & Speed, T. Replicated microarray data. *Statistica sinica* **12**, 31–46 (2002).

Chapter 9

Metagenomics

Sim Chun Hock,[a] Kee Shao Yong,[b] Ong Ai Ling,[b] Heng Huey Ying[b] and Teh Chee Keng[b]

[a]Faculty of Engineering, Computing and Science, Swinburne University of Technology, Kuching 93350, Sarawak, Malaysia
[b]Biotechnology & Breeding Department, Sime Darby Plantation R & D Centre, Selangor, 43400, Malaysia

Glossary of Terms

16s rRNA: This is a component of the 30S small subunit of prokaryotic (Archaea and bacteria) ribosomes. It is suitable for genetic diversity studies due to slow rates of evolution in some parts of the gene.

BIOM: This is a type of file format designed to be a standard format for representing biological sample by observation contingency tables. For more information, visit www.biom-format.org.

Operating taxonomic unit (OTU): A cluster of closely related organisms that have similar target gene sequences, normally at 97% identity.

Amplicon sequence variant (ASV): Single DNA sequence, which has a higher resolution than OTU, carry as least one nucleotide difference.

Introduction

Metagenomics is defined as sequence analyses of the total genomic DNA from environmental samples. The complex genomics information is useful to understand what microbial communities and their roles are present in the open ocean, soils, and even in the human gut. However, this has posed challenges to bioinformaticians to analyze and interpret. There are two general approaches for metagenomics studies; shotgun and targeted metagenomics.

Shotgun metagenomics is a study of random DNA/genes sequences which are mainly available in an assayed microbial community. With this method, studies on functional composition and biodiversity of known and unknown microbiota are now feasible. The main purpose is to study the functional composition of known and unknown organisms in the microbial community. Furthermore, shotgun metagenomics can also provide biodiversity of microbial communities. On the other hand, targeted metagenomics is a faster and cheaper way to obtain a microbial community/taxonomic profile. Targeted metagenomics can be defined as sequencing of targeted genes using polymerase chain reaction (PCR) using gene-specific primers. In metagenomics, most researchers focus on bacteria/archaea by targeting at least one or more highly variable regions of the 16S rRNA gene.[1] This gene is part of bacterial ribosomes, which contains conserved as well as variable sequences. The highly variable sequence region can be used as a molecular fingerprint marker to identify which taxa bacteria belong to. For other groups of organisms, different target genes are used. For example, internal transcribed spacer (ITS) is used for fungi[2] and 18S rRNA gene fragment is used for eukaryotes.[3]

In this practical, we use the MG-RAST[4–6] (https://www.mg-rast.org/) metagenomic analysis server to analyze shotgun metagenomic dataset. The MG-RAST is an open submission data server for processing, analyzing, sharing, and disseminating metagenomic datasets. In fact, it is a fully automated open-source server. The system is hosted in Argonne National Laboratory, Mathematics and Computer Science Division, Argonne, IL, USA since 2008.[4] It uses the M5 non-redundant protein database (M5NR) for functional annotation and M5 non-redundant taxonomy database (M5RNA) for taxonomic analysis. Data sources for M5NR are from European Bioinformatics Institute (EBI), Gene Ontology (GO), Join Genome

Institute (JGI), Kyoto Encyclopedia of Genes and Genomes (KEGG), National Center for Biotechnology Information (NCBI), Phage Annotation Tools and Methods (Phantome), The SEED Project (SEED), UniProt Knowledgebase (UniProt), Virginia Bioinformatics Institute (VBI) and Evolutionary Genealogy of genes: Non-supervised Orthologous Groups (eggNOG).[5] Data source for M5RNA are from Silva, Greengenes and RDP. Comparisons of metagenomic datasets with the M5NR or M5RNA database is a computationally intensive task as it involves phylogenetic comparisons, functional annotations, binning of sequences, phylogenomic profiling, and metabolic reconstructions.[7]

We use QIIME 2[8] to analyze targeted metagenomic dataset. The QIIME 2 is a rewrite of QIIME 1 to analysis next generation of microbiome data. With QIIME 2 plugin q2-feature-classifier,[9] we can perform taxonomic classification using any pre-trained classifiers in scikit-learn, such as classifier trained from Silva or Greengenes databases. The QIIME 2 also comes with a service called QIIME 2 View (https://view.qiime2.org/), which allows users to securely share and interact with microbiome results without installing the software.

Introduction to MG-RAST Server Workflow

Registration to MG-RAST

A registration is needed in order to kickstart analysis at the MG-RAST server (https://www.mg-rast.org/mgmain.html?mgpage=register) (Figure 1). Submission of sequence files to the MG-RAST for analysis is only possible, upon registration confirmation from the MG-RAST server. If you have registered an account, you may proceed to login.

Submission of Dataset

Firstly, we need to upload a sequence file to the MG-RAST server. Currently, the MG-RAST server supports shotgun and targeted (amplicon) metagenomic data from any sequencing platforms (e.g.

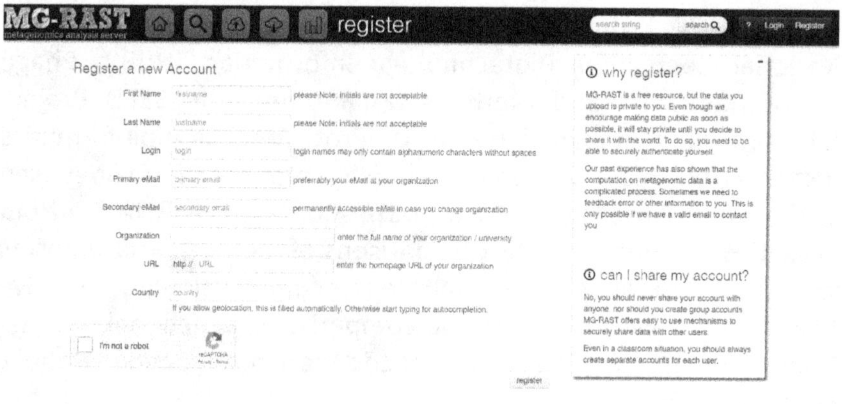

Figure 1. Registration of MG-RAST new user at https://www.mg-rast.org/mgmain.html?mgpage=register.

454 pyrosequencing, Illumina sequencing, and SOLiD sequencing) in FASTQ, FASTA, or SFF format. For pooled sequencing (two or more DNA samples pooled in a single sequencing run), we need to provide a barcode file for demultiplexing purpose. The barcode file should be in a plain text ASCII, containing lines with a barcode sequence followed by a unique filename separated by a tab. The MG-RAST demultiplexes by detecting the barcode region of each read. For example, if you have a sequence file testseq.fasta and your barcode file has tab-separated lines like:

```
AAAAAAAA    fileA
CCCCCCCC    fileC
```

The demultiplexing step will split your sequence file into three files:

`fileA.fasta containing all reads that begin with AAAAAAAA, fileC.fasta containing all reads that begin with CCCCCCCC, and testseq_no_MID_tag.fasta containing reads which do not match either of these two.`

After uploading the sequence file, it is a good practice to supply metadata for all metagenomics projects. Metadata include information about the project name, a detailed description of the isolation source, and scope of the project. The MG-RAST uses questionnaires to capture metadata for each project.

Job Status Monitor

Users may view the progress of their submitted job(s). The job is displayed in a table with sortable and searchable columns. For each job, an overview of progress is shown in a table with a series of color dots (green = completed tasks, blue = tasks being computed on, orange = next task to be queued, grey = pending for completion of another task they depend on, and red = an error). Figure 2 shows a workflow for the MG-RAST server and Figure 3 shows the summary of MG-RAST analysis pipeline for shotgun metagenomic sequences.

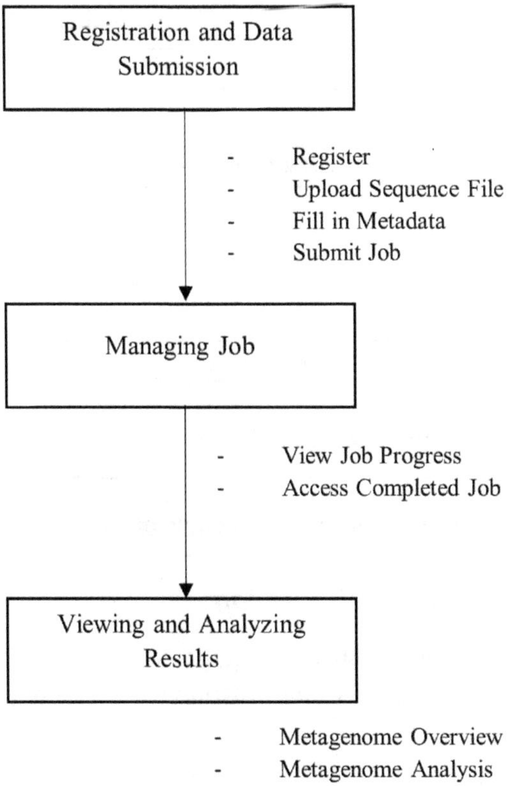

Figure 2. A workflow of the MG-RAST server.

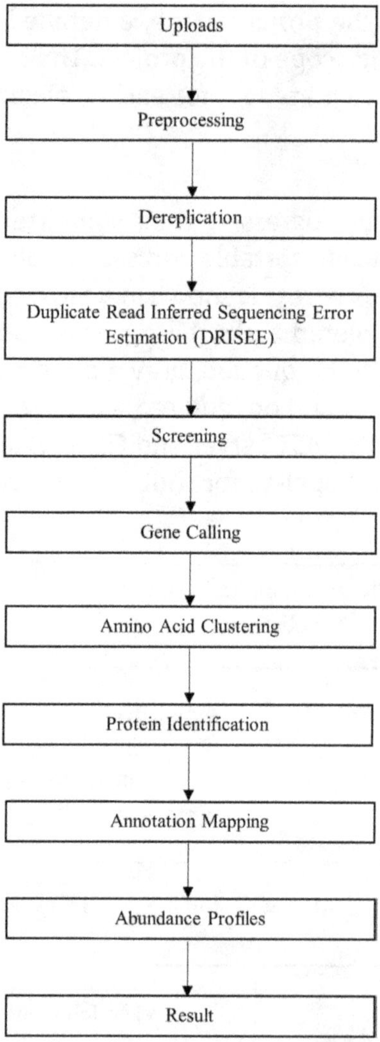

Figure 3. An analysis pipeline for shotgun metagenomic sequences.

Data Analysis and Result Viewing

For data analysis and result viewing, go to “Browse” page from the MG-RAST home page. This page displays a summary of submitted data and projects.

Analysis of Shotgun Metagenomic Sequence Datasets

In this practical, we used sequence files generated from Crohn's Disease Viral and Microbial Metagenome Project[10] (ERP001706). Six shotgun sequence files, which comprise of three representative fecal samples of patients with Crohn's disease and healthy volunteers (as negative controls), respectively. At the end of data analysis, we compared samples from Crohn's patients and healthy volunteers.

Getting Started

To download the sample data, visit https://www.ebi.ac.uk/ena/browser/view/PRJEB3206?show=reads. In the browser, tick on checkbox for required samples to download under column Generated FASTQ files (ftp). Then, click on "Download selected files" button. After completion, unzip the downloaded folder that consists of six selected shotgun sequence files in compressed FASTQ format (fastq.gz): ERR162917, ERR162919, ERR162921, ERR162933, ERR162935, and ERR162937 as shown in Table 1.

Table 1. Summary of sequence files downloaded from Crohn's Disease Viral and Microbial Metagenome Project at https://www.ebi.ac.uk/ena/browser/view/PRJEB3206?show=reads.

Run ID	Sample Name	Status	File Name	Size (MB)	Sequence Count	Sequence Type
ERR162917	C8	Crohn's disease	ERR162917.fastq.gz	57.2	62,064	454 WGS
ERR162919	C9	Crohn's disease	ERR162919.fastq.gz	20.7	26,145	454 WGS
ERR162921	C10	Crohn's disease	ERR162921.fastq.gz	22.1	29,035	454 WGS
ERR162933	V5	Healthy control	ERR162933.fastq.gz	12.0	18,556	454 WGS
ERR162935	V6	Healthy control	ERR162935.fastq.gz	9.8	18,620	454 WGS
ERR162937	V7	Healthy control	ERR162937.fastq.gz	8.4	12,707	454 WGS

Uploading and Submission

In the MG-RAST upload page, select and upload all 6 sequence files (Figure 4). For analysis submissions to the MG-RAST server, key in the information stated in Table 2 for each subsection. After

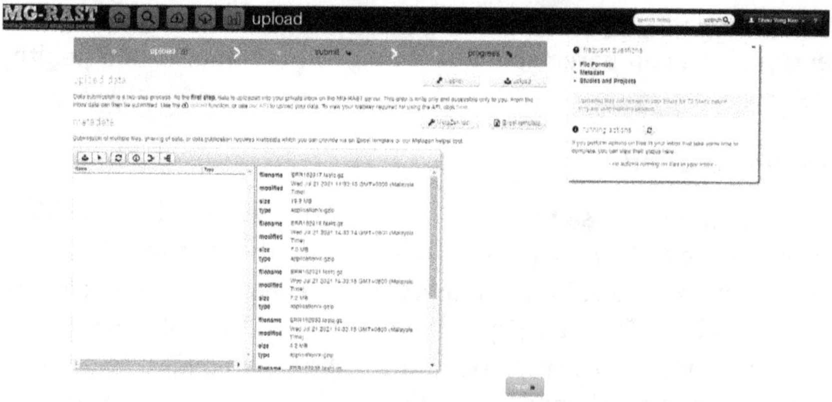

Figure 4. Uploading all input files to the MG-RAST server.

Table 2. MG-RAST submission of 6 compressed fastq files for shotgun metagenomic analysis. After keying in the information for every subsection, user must click "next" and make sure the subsection turns green.

	MG-RAST Submission	
Subsection	**Action**	**Remarks**
1. Select metadata file	Tick "I do not want to supply metadata"	For real metagenomic datasets, it is advisable to provide complete metadata information
2. Select project	Enter "meta_shotgun"	
3. Select sequence files(s)	Select the 6 fastq.gz files	
4. Choose pipeline options	Follow default setting	
5. Submit	Choose "Data will stay private (DEFAULT) — Lowest Priority" and submit job	Only applicable for this tutorial

submitting the analysis, a job number will be created automatically (Figure 5).

Results

To view results, go to "my studies" page from the MG-RAST home page. Click on the project name, "meta_shotgun". This leads to a new tab with project information and a table with all sample details (Figure 6). Click on the sample name, and a new tab, called

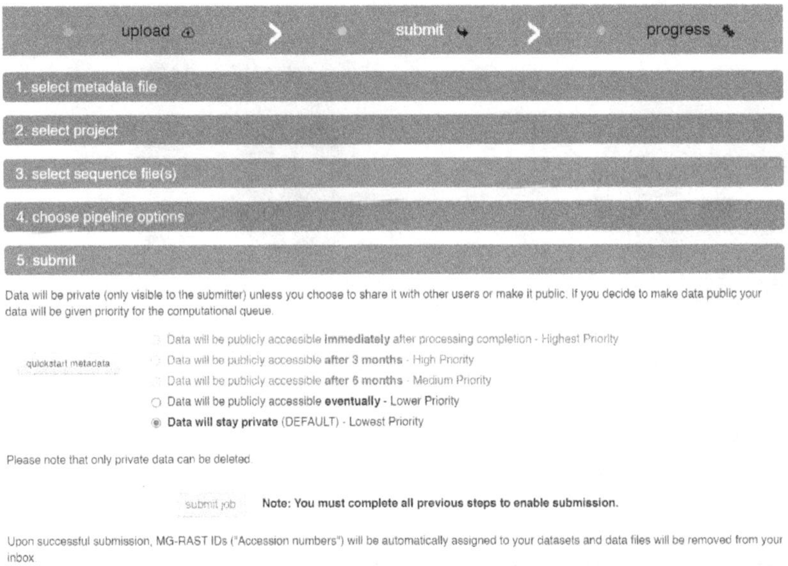

Figure 5. Submitting project to the MG-RAST server.

Figure 6. Summary information for metagenomes in project "meta_shotgun."

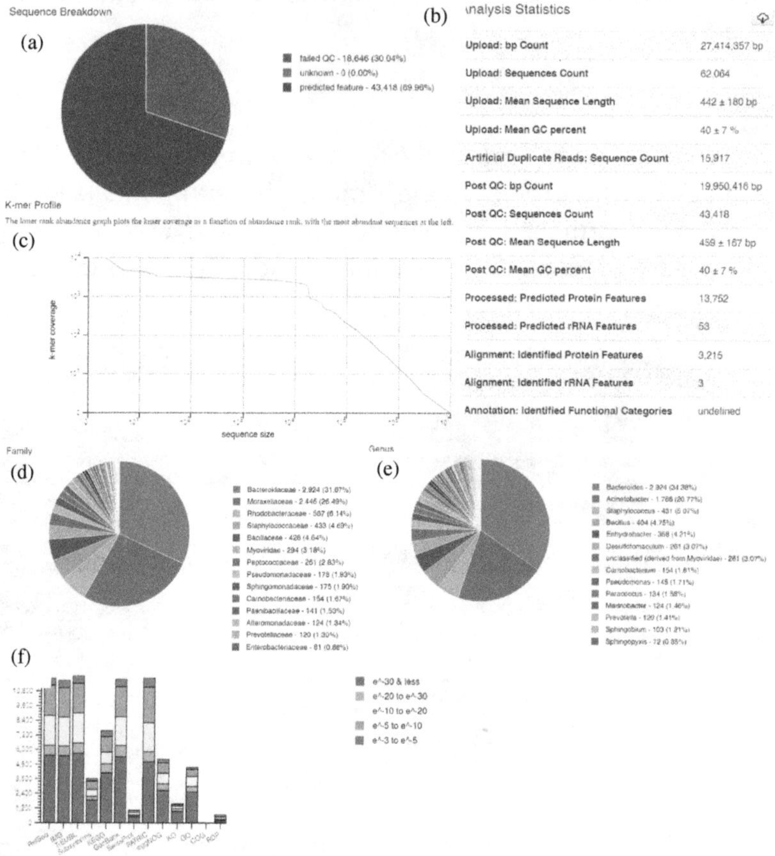

Figure 7. Outputs from the MG-RAST: (a) sequence breakdown, (b) analysis statistics, (c) k-mer curve, (d) family breakdown, (e) genus breakdown, (f) functional analysis breakdown in different databases for sample C9 (ERR162919).

“Metagenome Overview” appears with all results for respective sample name (Figure 6). Results for sample C9 (ERR162919) were presented in Figure 7.

A further comparative analysis between uploaded samples can be done in the MG-RAST via “Analysis” page. In the “Analysis” page, choose “RefSeq” and “COG” from the dropdown list under “available databases”, click on “Add”. Then, select the 6 shotgun

Figure 8. Create an analysis on RefSeq and COG databases.

sequences for analysis. Click on the green tick to allow analysis (Figure 8).

After processing is done, we can set parameters as shown in Figure 9 to filter the result. In this practical, subsequent analyses are performed at class level. Then, export the result as TSV or BIOM files. BIOM file can be loaded into MG-RAST or other analysis tools like QIIME 2 in future. Here, we export as TSV and open in Excel. Then, plot the relative abundance graph based on relative percentage against samples (Figure 10) and tabulate the predicted functions (Figure 11).

Figure 10 was derived from the shotgun metagenomic analysis and it reflected the dominance of class Gammaproteobacteria in control samples (V5, V6, and V7), while class Bacilli and Bacteroidia were observed to be more abundant in Crohn's disease samples (C8, C9, and C10).

Analysis of 16S rRNA-targeted Metagenomic Sequence Datasets

Getting Started

For the purpose of this practical, we use amplicon sequences representing the same sequence datasets from the Crohn's Disease Viral and Microbial Metagenome Project (ERP001706).[10]

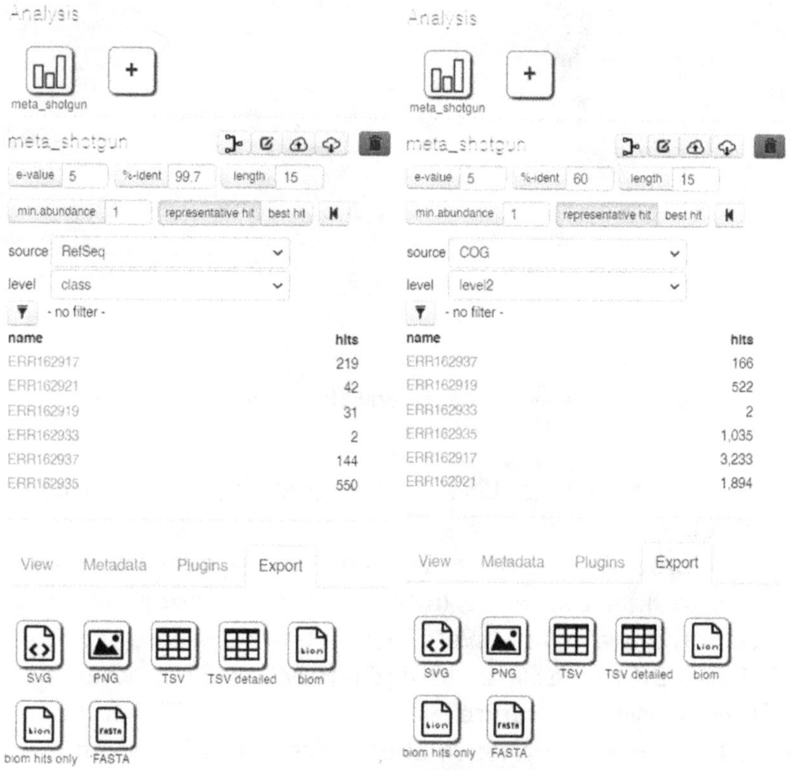

Figure 9. Analysis on RefSeq (left) and COG (right).

Visit https://www.ebi.ac.uk/ena/browser/view/PRJEB3206?show=reads to download the sample data. Then, tick the checkbox of samples listed in Table 3, under column "Generated FASTQ files (ftp)". Click on "Download selected files" button. After completion, unzip these downloaded files to FASTQ format.

We use the QIIME 2 for taxonomic classification of 16S rRNA-targeted metagenomic analysis (Figure 12). The QIIME 2 requires command-line knowledge, unlike the MG-RAST that has a user interface.

In this practical, we used Miniconda environment and wget command for QIIME 2 installation. In your terminal:

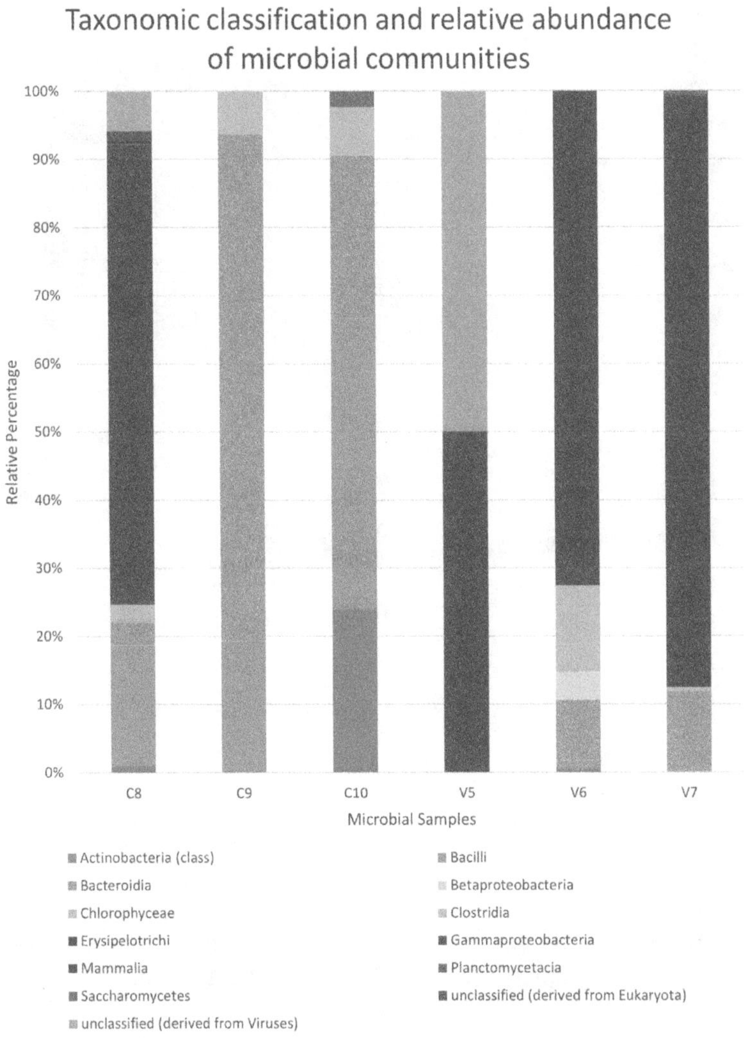

Figure 10. A taxonomic classification and relative abundance chart of microbial communities present in Crohn's Disease samples (C8, C9, C10) and in control samples (V5, V6, V7) generated in Excel. These microbial samples were classified up to class level from RefSeq database.

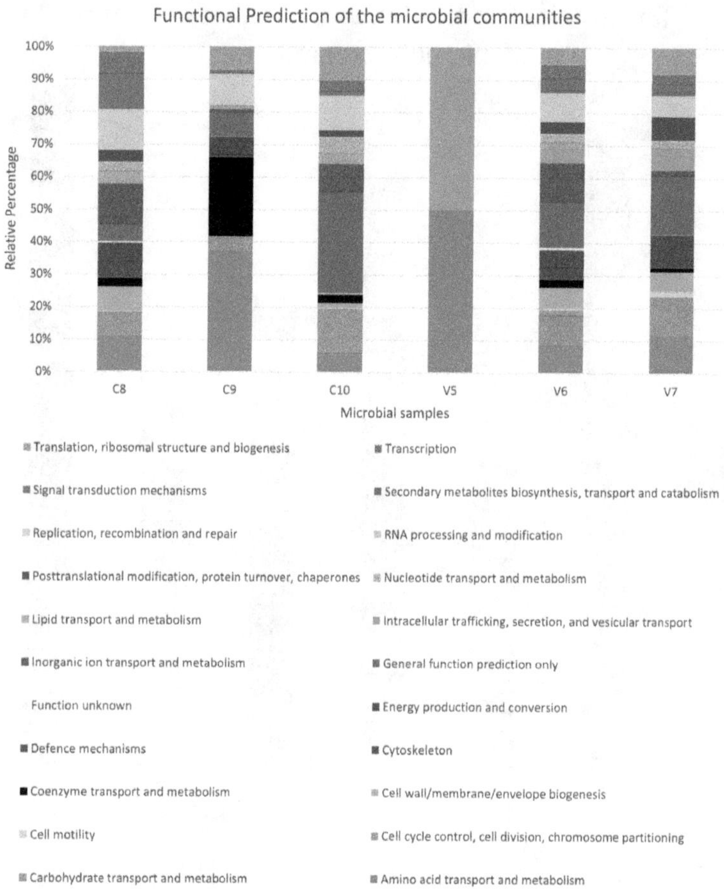

Figure 11. Functional prediction of microbial communities from COG present in Crohn's Disease samples (C8, C9, C10) and in control samples (V5, V6, V7).

Table 3. Summary of sample datasets of amplicon sequence downloaded from the Crohn's Disease Viral and Microbial Metagenome Project for 16S rRNA-targeted metagenomic analysis.

Run ID	Sample Name	Status	File Name	Size (MB)	Sequence Count	Sequence Type
ERR162918	C8	Crohn's disease	ERR162918.fastq.gz	6.4	6,968	454 Amplicon
ERR162920	C9	Crohn's disease	ERR162920.fastq.gz	8.5	10,863	454 Amplicon

Table 3. (*Continued*)

Run ID	Sample Name	Status	File Name	Size (MB)	Sequence Count	Sequence Type
ERR162922	C10	Crohn's disease	ERR162922.fastq.gz	7.1	9,278	454 Amplicon
ERR162934	V5	Healthy control	ERR162934.fastq.gz	7.6	8,205	454 Amplicon
ERR162936	V6	Healthy control	ERR162936.fastq.gz	10.9	12,110	454 Amplicon
ERR162938	V7	Healthy control	ERR162938.fastq.gz	8.2	9,031	454 Amplicon

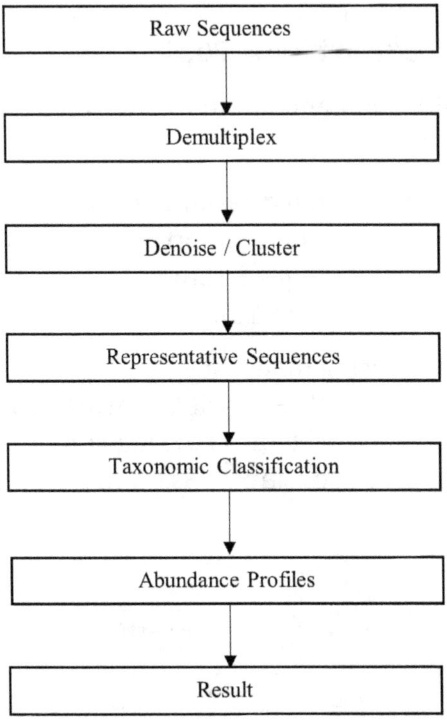

Figure 12. An analysis pipeline for 16S rRNA-targeted metagenomic sequences using QIIME 2.

```
# Make sure running latest version of conda
$ conda update conda

# Installing wget
$ conda install wget

# Install based on your OS platform. Here, we are using Linux-based environment
$ wget https://data.qiime2.org/distro/core/qiime2-2021.4-py38-linux-conda.yml
$ conda env create -n qiime2-2021.4 --file qiime2-2021.4-py38-linux-conda.yml

# Optional cleanup
$ rm qiime2-2021.4-py38-linux-conda.yml

# Activate the QIIME 2 environment in conda
$ conda activate qiime2-2021.4

# To test the installation and get more information of QIIME 2
$ qiime –help
```

In QIIME 2, all data must be imported and structured as an Artifact (.qza) format. We first import these raw sequences into a tab-separated file consisting of sample names and absolute file path as a manifest file (Figure 13).

```
# Import sequences using manifest
$ qiime tools import --type 'SampleData[SequencesWithQuality]' --input-path manifest --output-path 16s.qza --input-format SingleEndFastqManifestPhred33V2
```

```
sample-id	absolute-filepath
C8	../ERR162918.fastq.gz
C9	../ERR162920.fastq.gz
C10	../ERR162922.fastq.gz
V5	../ERR162934.fastq.gz
V6	../ERR162936.fastq.gz
V7	../ERR162938.fastq.gz
```

Figure 13. Example of the tab-separated file, named as "manifest" for this practical.

```
sample_name     run_id  genotype
#q2:types       categorical     categorical
C8      ERR162918       Crohn's disease
C9      ERR162920       Crohn's disease
C10     ERR162922       Crohn's disease
V5      ERR162934       Healthy control
V6      ERR162936       Healthy control
V7      ERR162938       Healthy control
```

Figure 14. Metadata used in this practical.

Then, denoise these sequences using DADA2 algorithms.[11] Subsequently, create a feature data consisting of counts associated with each sequence and feature. This step is required to create a metadata file as shown in Figure 14.

```
# Denoise sequences
$ qiime dada2 denoise-single \
  --i-demultiplexed-seqs ./16s.qza \
  --p-trunc-len 150 \
  --o-table ./dada2_table.qza \
  --o-representative-sequences ./dada2_rep_set.
qza \
  --o-denoising-stats ./dada2_stats.qza
```

```
# Create feature data, using metadata
$ qiime feature-table summarize \
  --i-table ./dada2_table.qza \
  --m-sample-metadata-file ./metadata.tsv \
  --o-visualization ./dada2_table.qzv
```

For this practical, we used the pre-trained naive Bayes machine-learning classifier, using Silva version 138 with 99% full-length OTU sequences to predict taxonomic classification of ASVs.[12,13] The pre-trained classifier can be downloaded from https://docs.qiime2.org/2021.4/data-resources/. To visualize the taxonomic composition that have been predicted, we need to build an interactive barplot of taxonomy for each sample.

```
# Taxonomic classification with pre-trained Silva classifier
$ qiime feature-classifier classify-sklearn \
  --i-reads ./dada2_rep_set.qza \
  --i-classifier ./silva-138-99-nb-classifier.qza \
```

```
  --o-classification ./silva_taxonomy.qza

# Build the interactive barplot of the taxonomy
$ qiime taxa barplot \
  --i-table ./dada2_table.qza\
  --i-taxonomy ./silva_taxonomy.qza \
  --m-metadata-file ./metadata.tsv \
  --o-visualization ./taxa_barplot.qzv
```

Visualization

QIIME 2 produces output files in .QZA and/or .QZV format. We could visualize our output files (silva_taxonomy.qza and taxa_barplot.qzv) using QIIME 2 View (https://view.qiime2.org/) interface by dragging these files from local machine to a dropbox provided (Figure 15).

The interactive barplot at class level (users can change taxonomic level) is displayed in Figures 16 and 17. Additionally, users can download the relative abundance of taxonomy into .CSV format and open it using Excel.

The taxonomic distribution of in Figure 17 reflects the dominance of Bacteroidia in Crohn's disease samples which is in

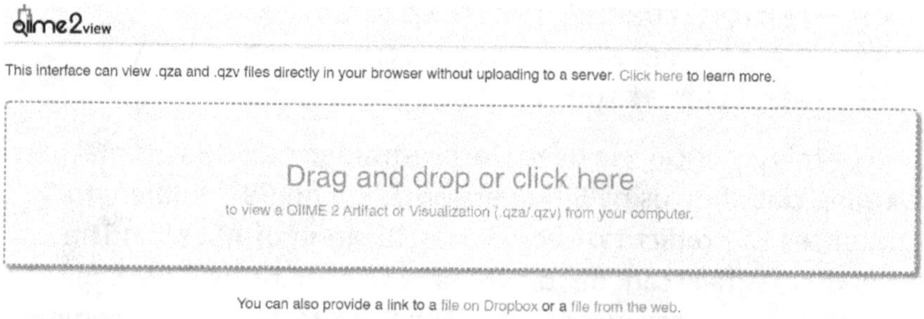

Figure 15. Interface of QIIME 2 View.

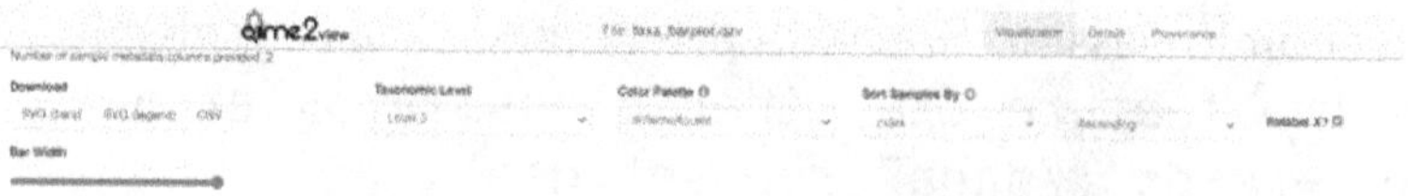

Figure 16. The barplot can be configured.

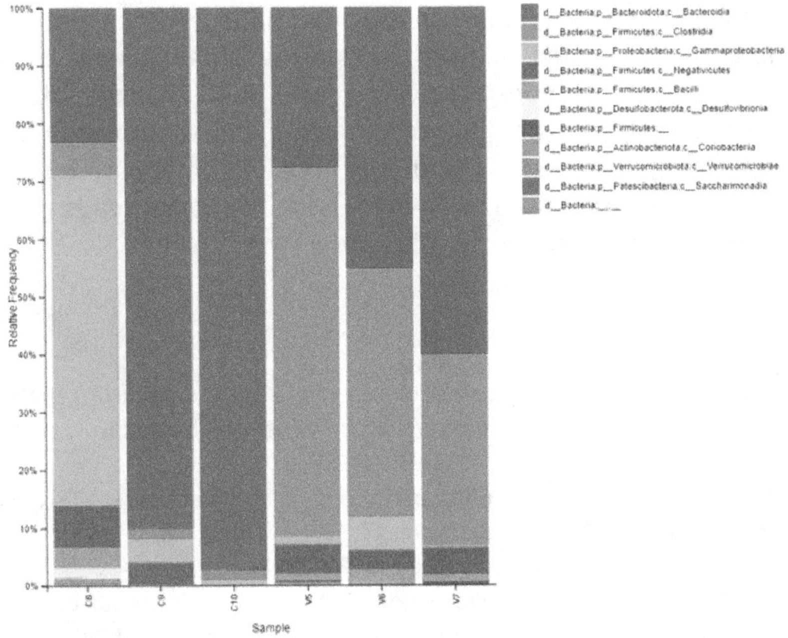

Figure 17. A taxonomic classification and relative abundance chart of microbial communities present in Crohn's Disease samples (C8, C9, C10) and in control samples (V5, V6, V7) generated in the QIIME 2, with the x-axis sorted by index. Microbial samples are classified up to class level (level 3).

agreement with the result of shotgun metagenomic analysis (Figure 10), while the distribution for control samples was found to be more diverse. This may suggest the importance in maintaining microbial diversity in human guts. The taxonomic classification in Figure 10 includes organisms other than Bacteria such as Eukaryote and Viruses, which cannot be found in Figure 17.

Conclusion

Each analysis method has its own pros and cons. Targeted metagenomic analysis is faster and cheaper, having more established pipelines for data analysis and equipped with more archived data as reference. However, targeted sequencing on specific genes only allows taxonomic classification of limited microbial groups, including bacteria and fungi. On the other hand, shotgun metagenomic dataset is obtained from sequencing broad regions of the genome, enabling more resolving power in detecting different organisms ranging from bacteria, virus, fungi to protozoa with more accurate taxonomic and functional annotations, but the method has higher false positive rates due to shorter reads generated from shotgun metagenomic sequencing.[14]

References

1. Klindworth, A. *et al.* Evaluation of general 16S ribosomal RNA gene PCR primers for classical and next-generation sequencing-based diversity studies. *Nucleic Acids Research* **41**, 1–11 (2013).
2. Schoch, C. L. *et al.* Nuclear ribosomal internal transcribed spacer (ITS) region as a universal DNA barcode marker for fungi. *Proceedings of the National Academy of Sciences, U. S. A.* **109**, 1–6 (2012).
3. Hadziavdic, K. *et al.* Characterization of the 18s rRNA gene for designing universal eukaryote specific primers. *PLoS One* **9**(2): p. e87624 (2014).
4. Meyer, F., Paarmann, D., D'Souza, M. *et al.* The metagenomics RAST server — a public resource for the automatic phylogenetic and functional analysis of metagenomes. *BMC Bioinformatics* **9**, 386 (2008). doi.org/10.1186/1471-2105-9-386.
5. Wilke, A. *et al.* The MG-RAST metagenomics database and portal in 2015. *Nucleic Acids Research* **44**, D59–D594 (2015).
6. Keegan, K. P., Glass, E. M. & Meyer, F. MG-RAST, a metagenomics service for analysis of microbial community structure and function. In: F. Martin, & S. Uroz (eds.), *Microbial Environmental Genomics (MEG), Methods in Molecular Biology*, 207–233. Springer Science+Business Media, New York, 2016, Vol. 1399.
7. Antonopoulos, D. A., Glass, E. M. & Meyer, F. Analyzing metagenomic data: inferring microbial community function with MG-RAST. In: R. W. Li (ed.), *Metagenomics and its Applications in Agriculture*, 47–60. Nova Science Publishers, 2010.
8. Bolyen, E., Rideout, J. R., Dillon, M. R. *et al.* Reproducible, interactive, scalable and extensible microbiome data science using QIIME 2. *Nature Biotechnology* **37**, 852–857 (2019). doi.org/10.1038/s41587-019-0209-9.

9. Bokulich, N. A., Kaehler, B. D., Rideout, J. R. *et al.* Optimizing taxonomic classification of marker-gene amplicon sequences with QIIME 2's q2-feature-classifier plugin. *Microbiome* **6**, 90 (2018). doi.org/10.1186/s40168-018-0470-z.
10. Pérez-Brocal, V. *et al.* Study of the viral and microbial communities associated with Crohn's disease: a metagenomic approach. *Clinical and Translational Gastroenterology* **4**, e36 (2013).
11. Callahan, B. J., McMurdie, P. J., Rosen, M. J., Han, A. W., Johnson, A. J., & Holmes, S. P. DADA2: High-resolution sample inference from Illumina amplicon data. *Nature Methods* **13**(7), 581–583 (2016). doi.org/10.1038/nmeth.3869.
12. Bokulich, N. A., Robeson, M., & Dillon, M. R. bokulich-lab/RESCRIPt. *Zenodo.* doi.org/10.5281/zenodo.3891931.
13. Bokulich, N. A., Kaehler, B. D., Rideout, J. R. *et al.* Optimizing taxonomic classification of marker-gene amplicon sequences with QIIME 2's q2-feature-classifier plugin. *Microbiome* **6**, 90 (2018). doi.org/10.1186/s40168-018-0470-z.
14. Ranjan, R., Rani, A., Metwally, A., McGee, H. S., & Perkins, D. L. Analysis of the microbiome: advantages of whole genome shotgun versus 16S amplicon sequencing. *Biochemical and Biophysical Research Communications* **469**(4), 967–977 (2016). doi.org/10.1016/j.bbrc.2015.12.083.

Chapter 10

Applications of NGS Data

Teh Chee Keng, Ong Ai Ling and Kwong Qi Bin

Biotechnology & Breeding Department, Sime Darby Plantation R&D Centre, Selangor, 43400, Malaysia.

Glossary of Terms

Locus: It is a genetic position in the genome and it can exist in a number of different allelic forms, which can often be traced as they are inherited by molecular or phenotypic markers.

Quantitative trait locus/loci (QTL): It is a section of DNA (at the locus) that correlates with variation for a quantitative trait (e.g. height and yield) that can vary in degree and be influenced by many genes and the environment.

Outcross population: This is derived from two genetically different parents, often producing full-sibs.

F_2 population: This is created by self-pollination of the F_1 derived from two different inbred parents or crossing between F_1 plants.

Backcross (BC) population: This is created by crossing an F_1 individual back to one of its parents.

Recombinant inbred line (RIL): This is created from inbreeding of individual lines of the F_2 generation and it generally requires 8 or more generations. Single seed descent can be used to speed up this process, with poor growing conditions leading to early flowering and limited seed set. A single seed for each line is taken to the next generation.

(*Continued*)

(*Continued*)

Population structure and cryptic relatedness: Population structure generally describes remote common ancestry of large groups of individuals, whereas cryptic relatedness refers to the recent common ancestry among smaller groups.

LOD: It stands for *logarithm (base 10) of odds* and it is used to estimate whether two genes, or two markers, or a marker and a qualitative phenotype, are likely to be located near each other on a chromosome and hence parental alleles are likely to be co-inherited.

Introduction

Rapid technological development in DNA sequencing has enabled the scientific community to sequence 418,755 organisms1 (as of October 2021) since the first genome of Bacteriophage MS2 was announced in 1976.[2] According to the Genome Online Database (GOLD), the number of genome projects shot up dramatically from 2012 and peaked at about 40,000 genomes per year in 2018 alone (Figure 1). The next question is how massive sequence data will benefit us. To address that, this chapter will discuss and illustrate some applications using NGS data to unveil the underlying biological mechanisms for a phenotype of interest, which is crucial in pharmacogenetics, agriculture and livestock research.

The genetic polymorphism in populations becomes important when mutated regions of the genome are discovered to influence phenotypic changes, such as susceptibility to diseases and increased crop yields which are no doubt exciting. To facilitate such discoveries, a reference genome sequence that links DNA markers to validated gene models, transcripts, proteins and other physical genomic features is required to better understand mutations behind phenotypic changes. By using NGS platforms, such as 454, ABI SOLiD, Illumina and Ion Torrent, short sequences of individual genomes can be mapped to a reference genome to reveal genetic polymorphism both within and between species. Some of these polymorphisms can

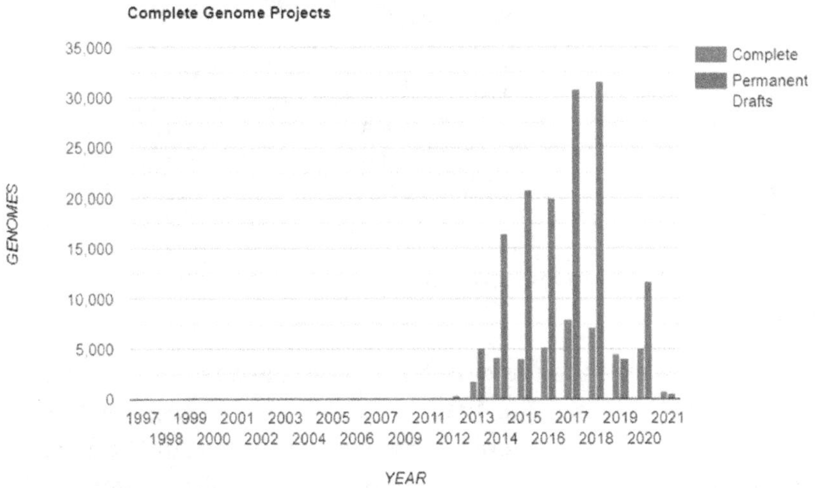

Figure 1. Complete and permanent draft genome totals in GOLD (by year and status).[1] Complete — complete genomes; Permanent drafts — draft genomes which are being updated.

be utilized as DNA markers, including random amplified polymorphic DNA (RAPD), restriction fragment length polymorphism (RFLP), amplified fragment length polymorphism (AFLP) and simple sequence repeat (SSR). SSR and single nucleotide polymorphism (SNP) are the most popular markers because of their high abundance, polymorphism, reproducibility and co-dominance.[3,4] To be more precise, SSRs are composed of short tandem arrays of simple nucleotide motifs and often have many allele forms at the same **locus**, whereas SNPs represents a single nucleotide change between individuals and tend to be bi-allelic (Figure 2). We have noted the preference shift from SSR towards SNP in recent years. Automation in high throughput assay formats up to 1,536 wells per plate has further made SNP marker analysis less laborious and more cost effective, compared to SSR marker. Also, the biallelic nature of SNP markers confers a much lower error rate in allele scoring, allowing higher levels of consistency between laboratories.[4]

With the help of DNA markers, the gene(s) of interest responsible for phenotype change can be identified and mapped onto the

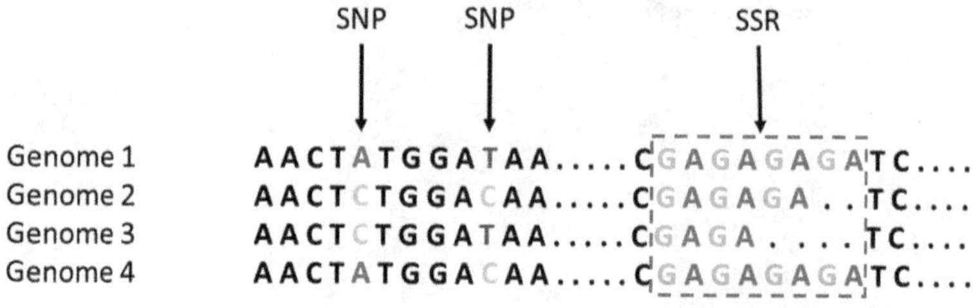

Figure 2. Single nucleotide polymorphism (SNP) and simple sequence repeat (SSR; boxed).

genome. Two methods of genetic mapping are commonly used to achieve this purpose. The first one is classical linkage analysis to determine the arrangement of markers or genes on the chromosomes based on meiotic recombination events within a family. The marker alleles that highly correlate to the phenotypic variation are expected to be close to genes influencing or controlling the phenotype. This correlation is defined as linkage, indicating two alleles of loci (between markers or between marker and phenotype) are co-inherited from parents. Nevertheless, the mapping resolution of the classical linkage method is always constrained by limited population size leading to insufficient recombination and sometimes by the lack of polymorphic markers. In humans, rare Mendelian diseases have been successfully localized through linkage mapping. However, inconsistent or ambiguous results for common non-Mendelian diseases (complex or quantitative phenotypes) are often reported.[5,6] In addition, human populations are often composed of small outbred families, rather than larger families possible with plants. Hence, human research groups began using an alternative method, which is association analysis. With the current sequencing and genotyping technologies, development of high density SNP panels is no longer a technical problem for most species. This has redefined the association strategy from an often candidate gene approach based on biochemical pathways to genome-wide association study (GWAS).[7] More importantly, association studies provide access to the total historical meiotic recombination events in a large heterogeneous population. With these, mapping resolution has improved close to the gene level. In 2003, the Human Genome Project (HGP) completed sequencing the

3 billion bp of human genome and it has become an important reference resource for the subsequent discovery of more than 1,800 disease-related genes. The agriculture and livestock research communities are following in steps of human studies.

This chapter will start from classical linkage mapping, assembly improvement based on a linkage map and GWAS. The aim is to provide some basic understanding of how to convert NGS data to valuable genetic information as mentioned above.

Classical Linkage Map

Classical linkage mapping was the first effort to determine the position of genetic factors affecting traits on chromosomes. The first linkage map based on morphological traits was constructed for fruit flies (*Drosophila melanogaster*) in year 1913,[8] which is 40 years earlier than the discovery of the molecular structure of DNA. The work successfully established the concept for genetic mapping. Alleles of loci (or genes or traits) in parental plants that co-locate on the same chromosome tend to be co-inherited and this is termed linkage. The pairwise distance between linked loci can be estimated according to numbers of meiotic crossovers observed, but this genetic distance does not reflect the physical distance (in bp). The distance is expressed in centimorgans (cM) which are calculated by applying a mapping function to the observed recombination frequency (number of observed recombinants/total number of observed recombinants; Rf), producing a linear distance. Pairs of polymorphic markers are compared in two point analysis to generate a network of Rf values between pairs of markers. If the Rf >= 0.5, the loci are considered unlinked. A large family that typically consists of 100 to 300 individuals with parental lines providing information on where alleles of each locus are coming from (also known as mapping population) is required to construct a good map using DNA markers. Without a reference genome, researchers rely on these linkage maps to locate **quantitative trait loci (QTL)** and gene regions for phenotypes of interest, such as linkage mapping of genes responsive to abiotic stress in barley[9], fatty acid compositions[10] and trunk height[11] in oil palm. Here, the R package *OneMap* is introduced for linkage mapping purposes.

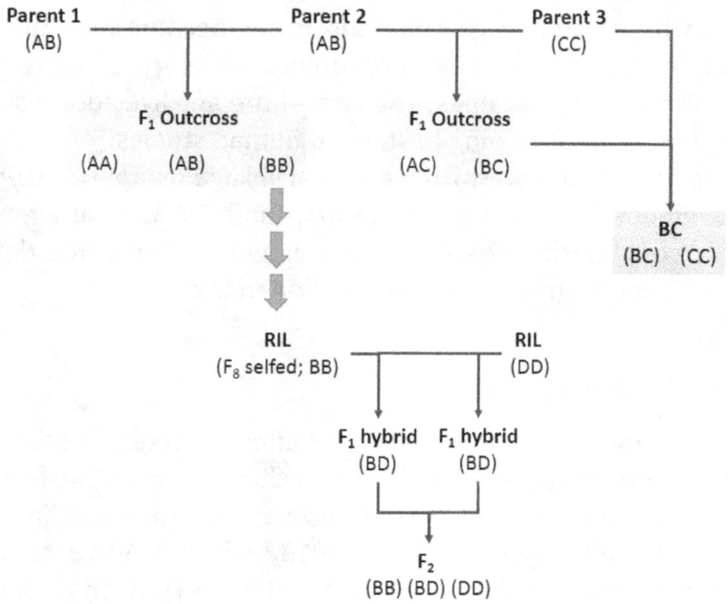

Figure 3. Experimental crosses. BC — backcross; RIL — recombinant inbred line.

OneMap (2.1.3)[12]

The software provides a platform for linkage map construction in various experimental crosses, including **recombinant inbred line (RIL)**, F_2, **backcross (BC)** and outcrossing populations (Figure 3). Outcross and **F_2 populations** confer higher mapping resolution, compared to BC populations. The latter population is less informative for linkage analysis because recombination is only observed among markers from one set of gametes from the donor parent (either male or female).[13] Thus, researchers are always advised to select the best experimental design based on available resources. In this section, a tutorial will be carried out in an **outcross population** using SNP markers. The SNP discovery and mining have been discussed in Chapter 7.

Installation

OneMap is an R package deposited at CRAN and can be automatically installed with the following command in R console:

```
>install.packages("onemap")
```

Alternatively, it can be downloaded using the command below:

```
$ wget"https://cran.r-project.org/src/contrib/
onemap_2.1.3.tar.gz"
```

Note: The package excludes other supportive packages that are required to be installed separately. Please refer to *OneMap* manual at CRAN for more information.

Input formatting

The input file is in text format, ...; the first line must indicates header with "data type xxxxx", where xxxxx representing the type of your mapping population, in this case we dealt with "outcross" population (Table 1). The second line indicates the number of individuals', 'number of markers', 'presence of CHROM data', 'presence of POS data' and 'number of traits'. These numbers must be separated with an empty space, you may also leave '0' for the last three items if you do not have the relevant information for them. The third line would be individual's name for the population separated by empty space. The genotype information is included separately for every marker. Each line of marker is started with asterisk "*". The software accepts multiallelic (e.g. SSRs), biallelic markers (SNPs) and also combinations of marker type. Alleles for each marker are differentiated based on "a", "b", "c" and "d". As mentioned earlier, we will work on SNP data, which have two alleles ("a" and "b") only. Thus, a reduced notation used to identify markers, cross types and genotypes in this section is given in Table 1 (Please refer to *OneMap* manual for more information).

Table 1. Reduced notation used to identify markers, cross types and genotypes.

Cross type	**Parent cross**	**Genotype segregation in offspring**	***Segregation ratio**
B3.7	ab × ab	aa, ab, bb	1:2:1
D1.10	ab × aa	aa, ab	2:1
D2.15	aa × ab	ab, aa	1:2

* The genotype segregation is expected to be compliant with Mendelian inheritance.

The informative markers for linkage analysis must be heterozygous in at least one of the parents (Recommendation: pre-determine the informative markers by genotyping the parents first).

Missing data are coded as "–"(minus sign) and a comma separates the information for each individual. An example input file for 10 individuals and 3 SNPs are given as follows. The input file must be saved in tab-delimited text format (".txt").

```
data type outcross
10 3 0 0 0
I1 I2 I3 (...total number of individuals)
*SNP1  B3.7     ab,ab,bb,aa,ab,aa,ab,-,bb,bb
*SNP2  D1.10    aa,aa,ab,-,ab,ab,aa,ab,ab,ab
*SNP3  D2.15    ab,ab,ab,aa,aa,aa,ab,-,aa,aa
```

Linkage mapping analysis

After the installation, the input file can be loaded to *OneMap* in the R console by:

Note: *OneMap* was built under R version 4.0.5 or more updated one.

```
>library(onemap)
#importing input data
>example.out<-read_onemap(inputfile="geno.input_recoded.raw")
```

The first step is to estimate the recombination fraction of all pairs of markers by using the default function (**LOD** score 5 and maximum recombination fraction 0.40) as:

```
>twopts<-rf_2pts(example.out)
#the LOD threshold and recombination fraction are adjustable.
>twopts<-rf_2pts(example.out,LOD=5,max.rf=0.4)
```

With the estimated recombination fraction and linkage phase for all pairs of markers, these markers could be assigned to different linkage groups (LG). The definition of LG is a network of marker pairs of which have shown linkage through two point analysis, therefore are likely to be on the same chromosome. Even though generating LGs is likely to provide significant information on the individual chromosomes, they are not necessary equivalent to LG. This is because many LGs can be found on the same chromosome due to low marker density and uneven recombination breakpoints. The function make.seq is used to specify which marker set that you want to analyze:

```
>mark.all<-make_seq(twopts,"all")
#to show the marker type
>marker_type(mark.all)
#to group the markers with adjusted LOD threshold and maximum rf
>LGs<-group(mark.all,LOD=5,max.rf=0.4)
>LGs
#to print the result of grouping
```

Within each LG, the mapping step can take place now. The mapping step is to determine marker order on the LG. The user must fix the mapping function i.e. Kosambi or Haldane as follows:

```
#to set Haldane's function
>set_map_fun("haldane")

#to set Kosambi's function (used in this section)
>set_map_fun("kosambi")
```

The user must then define which LG (Figure 4) is to be mapped. In this case, we only have LG1 and it is defined as:

```
>LG1<-make_seq(LGs,1)
```

Four two-point based algorithms, including, Rapid Chain Delineation, Recombination Counting and Ordering, and Unidirectional Growth can be used as below to order the markers.

```
> LGs
  This is an object of class 'group'
  It was generated from the object "mark.all"

  Criteria used to assign markers to groups:
    LOD = 5 , Maximum recombination fraction = 0.4

  No. markers:            25
  No. groups:             1
  No. linked markers:     25
  No. unlinked markers:   0

  Printing groups:
  Group 1 : 25 markers
    SNP-1 SNP-2 SNP-3 SNP-4 SNP-5 SNP-6 SNP-7 SNP-8 SNP-9 SNP-10 SNP-11 SNP-12 SNP-13 SNP-14 SNP-15 SNP-16 SNP-17 SNP-18 SNP-19 SNP-20 SNP-21 SNP-22 SNP-23 SNP-24 SNP-2
```

Figure 4. Screen capture of *OneMap* from the R console that shows the markers have been assigned to a single linkage group.

```
>LG1.rcd<-rcd(LG1)
>LG1.rec<-record(LG1)
>LG1.ug<-ug(LG1)
```

Note: A LG with less than or equal to 10 markers can be analyzed using a comparison of all functions:

```
>LG1.comp<-compare(LG1)
```

In this case, the Unidirectional Growth algorithm is used and the output is shown in Figure 5.

Eventually, the linkage map can be versualised in R Graphics (Figure 6), by the command:

```
>draw_map(LG1.ug,names=TRUE,cex.mrk=0.7)
```

Alternatively, there is a windows-based software (freely available) for visualizing maps, MapChart[14] at https://www.wur.nl/en/show/MapChart-2.32.htm.

The current version of MapChart is 2.32, with a simple user interface allowing the loading of marker ID and position in the 'Data' tab from the program (Figure 7a).

Only one LG is constructed using MapChart (2.32) for this practical illustration (Figure 7b). Nowadays, the development of a large number of DNA markers means there is no longer a bottleneck to construct a high-density linkage map at the genome-wide level. As an example, a genome-wide linkage map of oil palm (*Elaeis guineensis* Jacq.) which consists of 16 LGs, was constructed using 1,605 genic SNPs[15–16] (Figure 8). Ideally, the number of LGs should

```
> LG1.ug

Printing map:

Markers              Position           Parent 1       Parent 2

 1 SNP-1               0.00             a |   | b      a |   | b
 2 SNP-2               0.25             a |   | b      a |   | b
 3 SNP-3               1.26             a |   | b      a |   | b
 4 SNP-4               1.26             a |   | b      a |   | b
 5 SNP-5               1.75             a |   | b      a |   | b
 6 SNP-6               1.75             a |   | b      a |   | b
 7 SNP-7               2.75             a |   | b      a |   | b
 8 SNP-8               3.75             a |   | b      a |   | b
 9 SNP-9               4.75             a |   | b      a |   | b
10 SNP-10              5.00             a |   | b      a |   | b
11 SNP-11              5.75             a |   | b      a |   | b
12 SNP-12              6.00             a |   | b      a |   | b
13 SNP-13              6.50             a |   | b      a |   | b
14 SNP-14              6.75             a |   | b      a |   | b
15 SNP-15              7.75             a |   | b      a |   | b
16 SNP-16              8.25             a |   | b      a |   | b
17 SNP-17              8.25             a |   | b      a |   | b
18 SNP-18              8.25             a |   | b      a |   | b
19 SNP-19              9.75             a |   | b      a |   | b
20 SNP-20             10.00             a |   | b      a |   | b
21 SNP-21             11.76             a |   | b      a |   | b
22 SNP-22             11.76             a |   | b      a |   | b
23 SNP-23             11.76             a |   | b      a |   | b
24 SNP-24             13.80             a |   | b      a |   | a
25 SNP-25             15.81             a |   | b      a |   | a

25 markers            log-likelihood: -561.6452
```

Figure 5. Screen capture of linkage mapping and ordering for LG1. Marker name; linkage position (in cM) for each marker; Parent 1 and Parent 2 — parental genotypes.

be equalvalent to the number of chromosome pairs in a species (e.g. 23 pairs in human; 16 pairs in oil palm; 4 pairs in fruit fly), but this is not always true due to uneven distribution of recombination in the genome. The LG can split into fragments if low recombination frequencies occur in regions of the chromosome or if a long stretch of markers show no polymorphism, potentially due to their being identical by descent. In a study, a complete suppression of recombination in the centromeric and pericentromeric regions of

Figure 6. Illustration of a sample linkage map in R graphics.

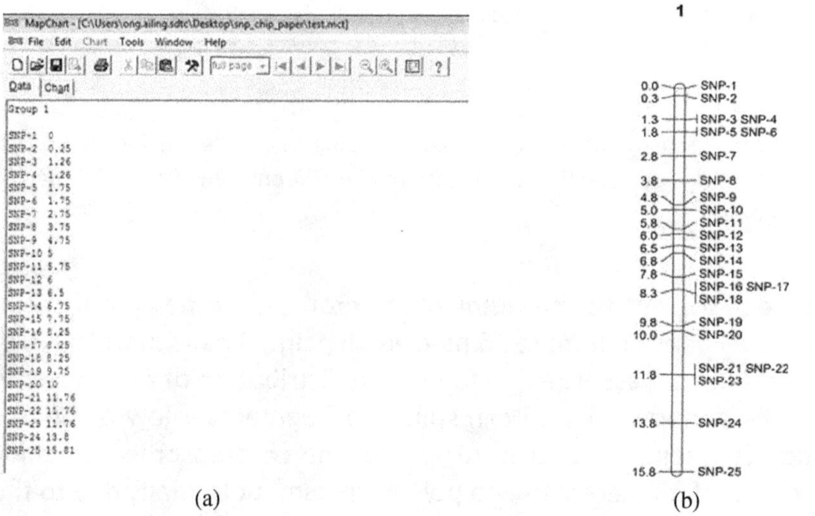

Figure 7. (a) Data input to MapChart (2.32); (b) Linkage map visualization for LG1.

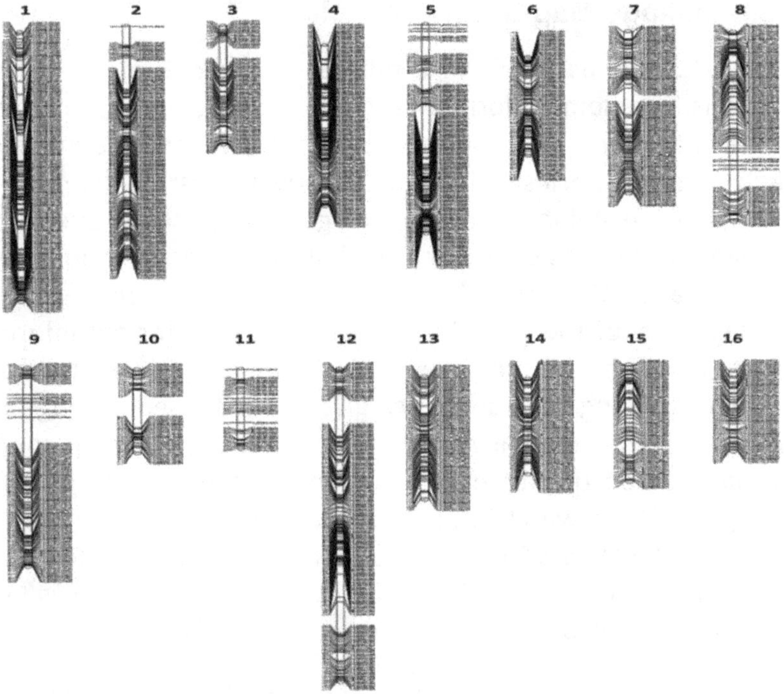

Figure 8. A genic SNP-based high density linkage map of a Deli *dura* x AVROS *pisifera* family. A total of 1,605 SNPs were assigned into 16 LGs with LOD threshold=4.0 and contributed to an average marker-marker interval of 0.8 cM.

papaya genome was identified.[15] In the same study, the long chromosome arm also showed a 60% higher recombination rate than the short arm.[17] Thus, if you have identified more LGs than the number of chromosome pairs of the species, this may be simply a reflection of recombination patterns or lack of polymorphism. Many linkage mapping programs, such as *OneMap* allow the users to adjust the LOD threshold. By increasing the threshold, the LG will tend to be split into smaller groups. Thus, the user is advised to start at a stringent LOD threshold which may lead to more groups than chromosome pairs, and subsequently try re-grouping at reduced stringency.[18] In other words, the determination of the number of LGs is not a straightforward task.

From Linkage Map to Physical Map

In the NGS era, assembly of a full genome is one of the central problems in genome informatics. A complete genome should ideally be in the form of chromosomes, as this will provide full information on the species of interest. However, most of the genome assemblies are still in the scaffold stage. This is mostly due to the complex nature of a genome, particularly in repetitive regions, centromeres and telomeres. Hence, raw reads which are essentially the fragments of the genome, often provide insufficient information for a full chromosome assembly. Some of the potential solutions include long read sequencing (e.g. PacBio sequencing) and optical mapping/sequencing. Another option is deploying combinatorial method of paired-end sequencing and DNA mapping to close gaps. Here, we illustrate the application of linkage map to anchor scaffolds as an effort to assemble the genome to give a physical map. The same linkage map with 25 SNP markers is used in this analysis according to a simple script as follows:

```
$ scaff2chr<LG cM input><scaffolds><cM to
kb rate><output>
```

<LG cM input>: The LG file, NGS_input_1_format.txt consists of linkage position (cM unit) of each marker, scafforld ID, physical position (bp) and scaffold length (bp) according to the format below (Figure 9).

<scaffolds>: A FASTA file, random_seq.fa contains scaffolds assembled using any type of *de novo* assembler, such as SOAPdenovo and Velvet.

<cM to kb rate>: The estimation is based on a simple regression graph of physical distance between markers within scaffold (bp) against linkage distance between markers (cM).

<output>: The output file is stored in FASTA format.

The regression graph in Figure 10 only consists of 16 points (marker pairs for each point). The remaining pairs are omitted from the analysis because these markers do not reside on the same scaforld. The physical distance thus, cannot be measured. In this case, the recombination rate is ~200 kb/cM. The value will be referred

SNPs_ID	cM	Scaffold ID	Position	Scaffold length
SNP-1	0.00	scaffold_spX_1	15730	490543
SNP-2	0.25	scaffold_spX_1	66422	490543
SNP-3	1.26	scaffold_spX_1	307133	490543
SNP-4	1.26	scaffold_spX_2	160380	595596
SNP-5	1.75	scaffold_spX_2	232906	595596
SNP-6	1.75	scaffold_spX_2	253654	595596
SNP-7	2.75	scaffold_spX_2	537549	595596
SNP-8	3.75	scaffold_spX_3	29359	270336
SNP-9	4.75	scaffold_spX_4	8775	253128
SNP-10	5.00	scaffold_spX_5	29558	194487
SNP-11	5.75	scaffold_spX_5	66458	194487
SNP-12	6.00	scaffold_spX_6	21022	260167
SNP-13	6.50	scaffold_spX_6	31802	260167
SNP-14	6.75	scaffold_spX_6	80770	260167
SNP-15	7.75	scaffold_spX_6	189272	260167
SNP-16	8.25	scaffold_spX_7	69621	437614
SNP-18	8.25	scaffold_spX_7	86037	437614
SNP-17	8.25	scaffold_spX_7	106723	437614
SNP-19	9.75	scaffold_spX_7	296432	437614
SNP-20	10.00	scaffold_spX_8	60790	378403
SNP-21	11.76	scaffold_spX_8	158399	378403
SNP-22	11.76	scaffold_spX_9	29131	301211
SNP-23	11.76	scaffold_spX_9	39813	301211
SNP-24	13.27	scaffold_spX_10	13896	224086
SNP-25	14.28	scaffold_spX_10	93641	224086

Figure 9. Sample input for scaf2chr.

to estimate the gap interval between scaffolds, which is important in gap closing.

The example command is executed as:

```
$ scaff2chr NGS_input_1_format random_seq.fa
200000>built.fa
```

The 25-linked SNP markers which initially located on 10 scaffolds, are successfully oriented and anchored as one pseudomolecule/chromosome sequence. The process is visualized in Figure 11 for a better understanding.

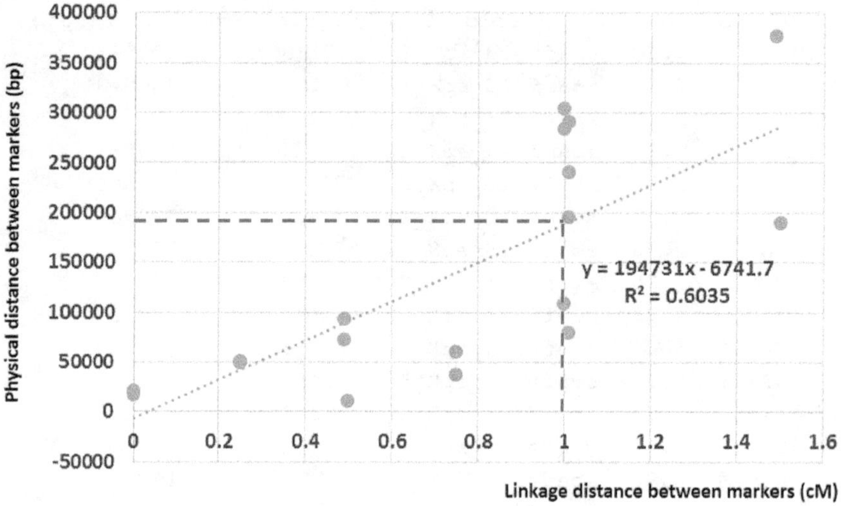

Figure 10. A regression graph of physical distance between markers against linkage distance between markers.

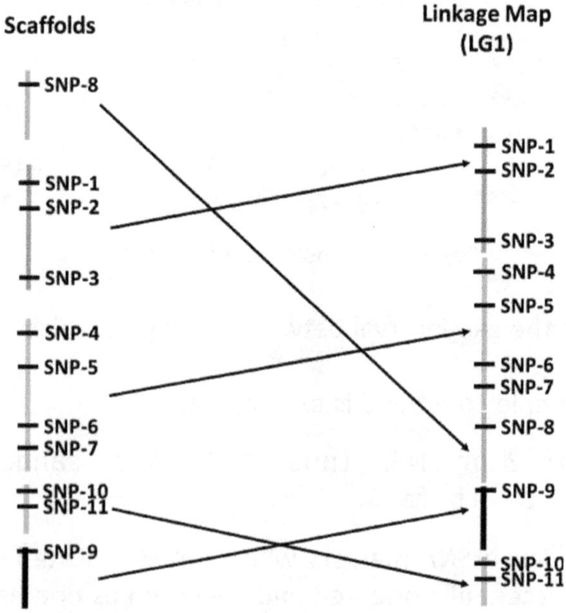

Figure 11. Scaffold anchoring based on linkage map (LG1). The 'Linkage Map' here is referred to as pseudomolecule/chromosome after the scaffold rearrangement.

In some cases, the linkage position of the marker is family specific because of recombination difference across families. To improve mapping accuracy, a concensus linkage map is usually generated by merging several maps, as reported in stickleback (*Gasterosteus aculeatus*)[19] and chicken.[20] Scaffold misalignment (especially the long ones) and recombination hotspots can be detected by comparing linkage and physical map positions. An example of such phenomena is given in Figure 12. Two ends of the chromosome 1 (telomere) have a higher recombination rate compared to the plateau around the middle (centromere). The good correlation between the genetic and physical map positions indicates a good quality of genome assembly, except that scaffolds reside within 70–80 Kb due to scaffold misarrangment. The same method has been used in many species, including human,[21] stickleback,[19] and cotton[22] to improve their reference genome.

Alternatively, genome assembly can be improved further using ALLMAPS[23] by integrating more than one linkage map. Multiple linkage maps enhance the confidence level of marker ordering and complement each other to expand the mapping coverage on a physical genome. The approach has been reported in oil palm assembly using three linkage maps to successfully assign

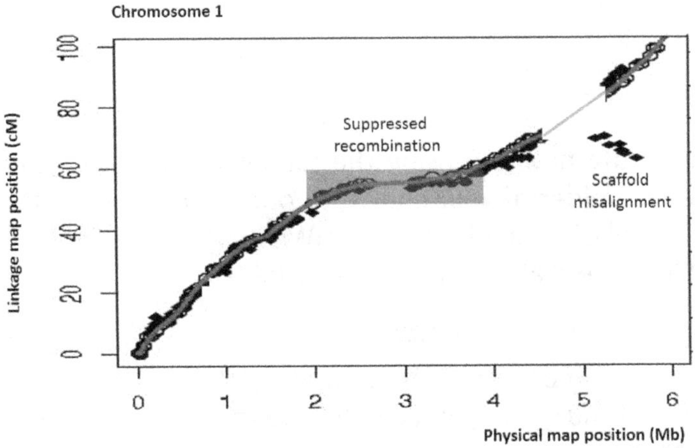

Figure 12: An example plot of genetic vs. physical map position from chromosome 1.

thousands of previously unplaced scaffolds into the reference genome[24]. Similar to Figure 12, high correlations between linkage and physical positions per chromosome were observed in this study, clearly distinguishing both telomere regions from centromeric region for each chromosome.

Genome-wide Association Studies (GWAS)

So far, we have discussed classical linkage mapping. The integration of linkage map and physical map can actually further improve the genome assembly quality. With a good reference map, we can move on to locate QTLs for phenotypes of interest on the genome through GWAS, which we believe, is one the most useful applications of a genome. As mentioned earlier, GWAS confers a much higher mapping resolution compared to classical linkage mapping. Here, we introduce *PLINK 1.90 beta* for the association analysis.

PLINK 1.90 Beta

PLINK[25] is an open-source whole genome association analysis toolset, designed for a range of basic, large-scale analyses, in a computationally efficient approach. Below is a simple tutorial for the program.

Installation

Download the program using the following command and unzip the file in a folder of your choice. In our example, we will be running the analysis in a folder called GWAS. Please take note that the input files should also be in the same folder.

```
$ mkdir GWAS; cd GWAS
$ wget https://s3.amazonaws.com/plink1-assets/
plink_linux_x86_64_20210606.zip --no-check-
certificate
```

```
$ unzip plink_linux_x86_64_20210606.zip
```

To run the plink program, just type the command './plink' in the same directory

Input files and format

The program requires two main input files i.e. PED and MAP. The PED file format consists of 6 mandatory columns as header while column 7 denotes the genotypes for each individuals for each markers:

Family ID

Individual ID

Paternal ID

Maternal ID

Sex (1=male; 2=female; other=unknown)

Phenotype (-9 is missing phenotype or in separate file)

The MAP file describes the information for each assayed marker and the file consists of 4 columns:

Chromosome

Marker identifier

Genetic distance (morgans)

Base-pair position (bp units)

Next would be the optional phenotype file for either quantitative or binary with the format of first two columns giving the family ID and individuals ID (just like the first two columns of PED file); the rest of the column is phenotypic data. The binary phenotype consists of two distinct traits (affected/unaffected or present/absence), whereas the quantitative phenotype is a measureable trait that depends on the accumulation of multiple genes and the environment.

Download datasets

Datasets can be downloaded at
https://github.com/PU-SDS/ngs-book-dataset/tree/master/chapter10

They comprise bi.phe, qt.phe, test_data.map and test_data.ped files.

Note: Make sure all files are stored at the same directory.

The example data consists of 311 samples with known pedigree and ambigous gender were genotyped using 25,018 SNP markers. The assayed samples are diploid with $2n = 2x = 5$ i.e. five homologous chromosome pairs in the genome. The same samples can have more than one phenotype or a combination of binary and quantitative phenotypes. In such cases, phenotypes need to be stored separately as binary data (bi.phe) and quantitative data (qt.phe). Both types of phenotypic data are provided in this section. The user is advised to convert the PED (together with MAP) to a binary file format which is more compact, so that the subsequent analysis can be expedited. To make the binary file, use the following command:

```
$ ./plink --file test_data --make-bed --out test_
data
```

Note: An output log file with data summary is generated.

This step will produce three files with similar prefixes, including (i) test_data.bim which consists of marker information, (ii) test_data.fam which denotes pedigree information for each individual, and (iii) test_data.bed which is the compressed binary file of genotypes.

Association analysis

The *PLINK* program is able to analyze binary and quantitative phenotypes. The simplest form of association analysis is a single marker test. To aid understanding of involved tests, we will go through two examples, one with binary phenotypes and another with quantitative phenotypes. For binary phenotypes, it is a chi-square test between case (affected) and control (unaffected) populations. In Figure 13a, a simple chi-square test with 1 degree of freedom (d.f.) was done to compare Allelic counts of SNP1 between a Diabetes mellitus (DM)-infected population and a control population. Usually, individuals in the case population and the control population are coded as '1' and '0', respectively in bi.phe file (Note: the binary code can also be other numbers, such as '1' and '2'). The result indicates that those individuals with G allele of SNP1 is significantly associated to DM (*p-value* = 1.8×10^{-4}). As for quantitative phenotypes, we adopt linear regression analysis (e.g. ANOVA) to measure whether there is any significant difference among samples. An example is given in Figure 13b. The SNP2 is significantly associated to height phenotype (*p-value* = 1.0×10^{-17}) and those individuals who carry G allele of SNP2 are averagely taller. The same analysis is eventually applied on every marker throughout the genome to perform GWAS.

Allele	Case (MD)	Control (unaffected)
G	526	446
T	474	554

(a)

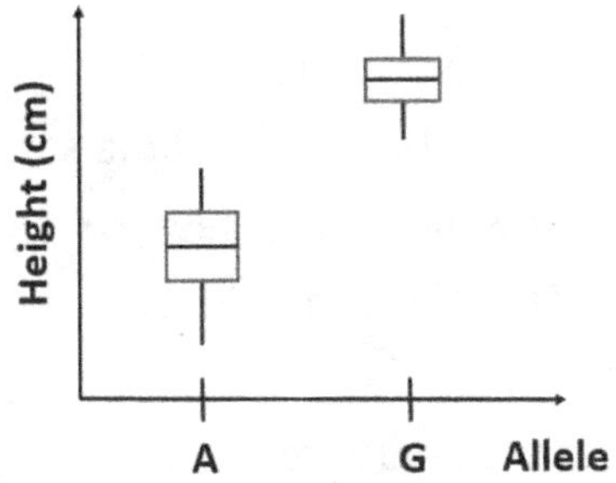

(b)

Figure 13. (a) A simple chi-square test (1 d.f.) on allele counts of SNP1 in Diabetes mellitus (DM)-infected population (1) and control population (0) with *p-value* = 1.8×10^{-4}; (b) Boxplots of height distribution (cm) grouped according to A allele and G allele of SNP2 with a linear regression, ANOVA *p-value* = 1.0×10^{-17}.

Take note that default association analysis is based on allelic model (i.e. 2 alleles per SNP with d.f. = 1). However, the d.f. can be more than 1 when genotypic model (co-dominant model with three genotypes per SNP) or multiallelic model is used. The alternative genetic models will be discussed later.

Firstly, a basic association test for the binary phenotype can be done as:

```
$ ./plink --bfile test_data --assoc --out basic_
test --pheno bi.phe --allow-no-sex
```

Flag:

--bfile: as the input file prefix
--assoc: to perform the association test
--pheno: as the phenotype input file (an example data is provided)
--out: as the output file prefix

Note: '--allow-no-sex' is added to disable the automatic setting of the phenotype to missing if the individual has an ambiguous sex code.

This will generates an output file, 'basic_test.assoc' with the column headings as shown below:

CHR	Chromosome
SNP	SNP ID
BP	Physical position (base-pair)
A1	Minor allele name (based on whole sample)
F_A	Frequency of this allele in cases
F_U	Frequency of this allele in controls
A2	Major allele name
CHISQ	Basic allelic test chi-square (1df)
P	Asymptotic p-value for this test
OR	Estimated odds ratio (for A1, i.e. A2 is reference)

For quantitative traits, use the following command:

```
$ ./plink --bfile test_data --assoc --out basic_
qt --pheno qt.phe --allow-no-sex
```

This will generate an output file, 'basic_qt.qassoc' with column headings as shown below:

CHR	Chromosome number
SNP	SNP identifier
BP	Physical position (base-pair)
NMISS	Number of non-missing genotypes
BETA	Regression coefficient
SE	Standard error
R2	Regression r-squared
T	Wald test (based on t-distribtion)
P	Wald test asymptotic p-value

Only the association output for the binary phenotype is further used for subsequent analyses and interpretations. The GWAS is known to be susceptible to confounding factors, particularly if **population structure and cryptic relatedness** exist in assayed samples (or discovery population).[26] These confounding factors here produce inflated false positives, which relate to the distribution of genotypes between sub-structures instead of accounting for the phenotypic variance. The *PLINK* program confers various correction models (with different stringencies) to address the confounding factors. To perform these, we can rerun the association analysis, adding the '--adjust' flag as shown below.

```
$ ./plink --bfile test_data --assoc --out basic_
test --pheno bi.phe --adjust --allow-no-sex
```

Flag:

--adjust: to generate multiple testing corrected *p-value*

As mentioned, the '--adjust' function includes various correction models. This will generate an output file, 'basic_test.assoc.adjusted' with column headings as shown in Figure 14.

CHR	Chromosome
SNP	SNP identifier

CHR	SNP	UNADJ	GC	BONF	HOLM	SIDAK_SS	SIDAK_SD	FDR_BH	FDR_BY
3	snp44196	3.288e-13	1.743e-05	8.163e-09	8.163e-09	8.163e-09	8.163e-09	8.163e-09	8.731e-08
3	snp79910	1.141e-10	0.000143	2.833e-06	2.833e-06	2.833e-06	2.833e-06	1.416e-06	1.515e-05
3	snp51552	5.313e-10	0.0002494	1.319e-05	1.319e-05	1.319e-05	1.319e-05	4.397e-06	4.703e-05
3	snp79911	2.961e-09	0.000465	7.352e-05	7.351e-05	7.352e-05	7.351e-05	1.487e-05	0.0001591
3	snp61021	3.096e-09	0.0004726	7.686e-05	7.685e-05	7.686e-05	7.684e-05	1.487e-05	0.0001591
3	snp62103	3.898e-09	0.0005138	9.677e-05	9.675e-05	9.676e-05	9.674e-05	1.487e-05	0.0001591
3	snp46393	4.793e-09	0.0005539	0.000119	0.0001189	0.000119	0.0001189	1.487e-05	0.0001591
3	snp46394	4.793e-09	0.0005539	0.000119	0.0001189	0.000119	0.0001189	1.487e-05	0.0001591
3	snp69961	1.097e-08	0.0007486	0.0002723	0.0002722	0.0002722	0.0002721	2.862e-05	0.0003061

Figure 14. Association analysis with an adjusted *p-value* using different correction models.

UNADJ	Unadjusted p-value
GC	Genomic control[27] adjusted p-value
BONF	Bonferroni adjusted p-value
HOLM	Holm step-down adjusted significance value
SIDAK_SS	Sidak single-step adjusted significance value
SIDAK_SD	Sidak step-down adjusted significance value
FDR_BH	Benjamini & Hochberg (1995) step-up FDR control
FDR_BY	Benjamini & Yekutieli (2001) step-up FDR control

The genomic inflation factor estimated lambda (as GIF) is defined as the ratio of the empirically observed median chi-squared distribution of the test statistic (*p-value* of SNP markers) to the expected median, so the extent of the bulk inflation and excess false positive signals can be quantified. In an output log file (Figure 15), the GIF without correction is 2.87399, indicating an inflated positive result due to population structure in the dataset. The optimal GIF should be close to 1.0 under the null hypothesis. GIF = 1.0 indicates that the observed *p-value* distribution equals to the expected distribution. This, however, explains no significant association signals detected. Hence, a good GIF should be more than 1.0, but lower than 1.1, if possible.

Take note that the GIF after GC is not given in the log file, but we can estimate the GIF based on the *p-values* using an R command as below. To run this code, first initiate R by typing "R" in the command prompt.

```
PLINK v1.90p 64-bit (2 Oct 2015)
Options in effect:
  --adjust
  --allow-no-sex
  --assoc
  --bfile test_data
  --out basic_test
  --pheno bi.phe

Hostname: SDTC
Working directory:
Start time: Fri May 13 15:15:28 2016

Random number seed: 1463123728
9881 MB RAM detected; reserving 4940 MB for main workspace.
25018 variants loaded from .bim file.
311 people (0 males, 0 females, 311 ambiguous) loaded from .fam.
Ambiguous sex IDs written to basic_test.nosex .
310 phenotype values present after --pheno.
Using 1 thread (no multithreaded calculations invoked.
Before main variant filters, 311 founders and 0 nonfounders present.
Calculating allele frequencies... done.
25018 variants and 311 people pass filters and QC.
Among remaining phenotypes, 214 are cases and 96 are controls.  (1 phenotype is
missing.)
Writing C/C --assoc report to basic_test.assoc ... done.
--adjust: Genomic inflation est. lambda (based on median chisq) = 2.87399.
--adjust values (24825 variants) written to basic_test.assoc.adjusted .
```

Figure 15. An output log file of association analysis with correction methods ('--adjust').

```
## calculates lambda by the median method for GC model
>S <- read.table("basic_test.assoc.adjusted",header=T)
>data<-S[,"GC"] #p-value column
>data<- qchisq(data, 1, lower.tail = FALSE)
>median(data, na.rm = TRUE)/qchisq(0.5, 1)
1.002335
```

The same script can be repeated with *p*-values of each correction model (BONF, FDR_BH, HOLM etc.) to estimate their GIF values. In this case, we have selected GC as the correction method. By using this method, the population stratification is successfully addressed in the association result when GIF declines to 1.002335.

All analyses above were done using the basic allelic model, which compares allelic frequencies between cases and controls. The program is also able to perform association analyses other than the basic allelic test. These options can be accessed by using '--model' function, including:

1. Cochran-Armitage trend test
2. Genotypic test (co-dominant)
3. Dominant gene action test
4. Recessive gene action test

Each of these models makes different assumptions about the input data. Unlike the basic allelic test, the Cochran-Armitage trend test does not assume Hardy-Weinberg equilibrium (HWE). The individual, not the allele, is the unit of analysis. This feature is to retain those markers with severe deviations from HWE. In many cases, these deviations reflect population stratification in samples or bad marker quality. However, this is not always true, because some of these markers can be genuine due to selection pressure. Another model uses genotypes instead of alleles. This is particularly useful since association could be due to co-dominant (genotypic model) or dominant-recessive effects of the minor allele (the minor alleles could be found in the output of either the '--assoc' or the '--freq' functions). Presuming D is the minor allele, while d is the major one. The allele assignment for the tests are stated as follows:

Allelic:	D	versus		d	
Dominant:	(DD, Dd)	versus		dd	
Recessive:	DD	versus		(Dd, dd)	
Genotypic:	DD	versus	Dd	versus	dd

The command for this analysis is:

```
$ ./plink --bfile test_data --model --out mod
--allow-no-sex --pheno bi.phe --snp snp88763
```

Flag:

--snp: SNP selected for association analysis, ignore if the plan is to run for all markers

An example output with column headings is given in Figure 16.

CHR	Chromosome
SNP	SNP identifier

CHR	SNP	A1	A2	TEST	AFF	UNAFF	CHISQ	DF	P
1	snp88763	1	2	GENO	48/106/60	12/61/23	6.202	2	0.045
1	snp88763	1	2	TREND	202/226	85/107	0.4977	1	0.4805
1	snp88763	1	2	ALLELIC	202/226	85/107	0.4562	1	0.4994
1	snp88763	1	2	DOM	154/60	73/23	0.5624	1	0.4533
1	snp88763	1	2	REC	48/166	12/84	4.186	1	0.04075

Figure 16. An example output of an association analysis using different genetic models.

A1 & A2	Allele 1 & Allele 2
TEST	Type of test (using different genetic model)
ALLELIC	Basic allelic test
TREND	Cochran–Armitage trend test
GENO	Genotypic test
DOM	Dominant model
REC	Recessive model
AFF	Allelic/Genotypic frequency of affected (Case)
UNAFF	Allelic/Genotypic frequency of unaffected (Control)
CHISQ	Chi-Sq test statistic
DF	Degrees of freedom
P	P-value

In this case, snp88763 marker with $p = 0.045$ indicates a significant result under the genotypic test (at threshold of $p>0.05$), but not in other genetic models.

As you can see, the *PLINK* program provides all GWAS outputs in text format only. Indeed, we can visualize the output as Manhattan plots using an R package 'qqman',[28] which is available online at http://cran.r-project.org/web/packages/qqman/. Firstly, initiate R as usual:

```
>install.packages("qqman")
>library(qqman)
#the same uncorrected association output
>results<-read.table("basic_test.assoc",header=T)
>results <- results[,c("SNP", "CHR", "BP", "P")]
```

```
#must define the headers accordingly
> results<-na.omit(results)
>manhattan(results)
```

The uncorrected GWAS output based on the 'basic_test.assoc' generated in the *PLINK* program is shown as a Manhattan plot (Figure 17), which is the common way to present output in many GWAS publications. Genomic positions are indicated on the X-axis, whereas the negative logarithm of the association *p-value* ($-\log_{10}$ (p)) for each SNP marker is displayed on the Y-axis. Highly associated markers have the smallest *p-value,* but their $-\log_{10}$ (p) will be the greatest. The uncorrected GWAS output with GIF = 2.87399 showed inflated false positive in samples, which is also reflected in Figure 17. Too many association signals are detected based on

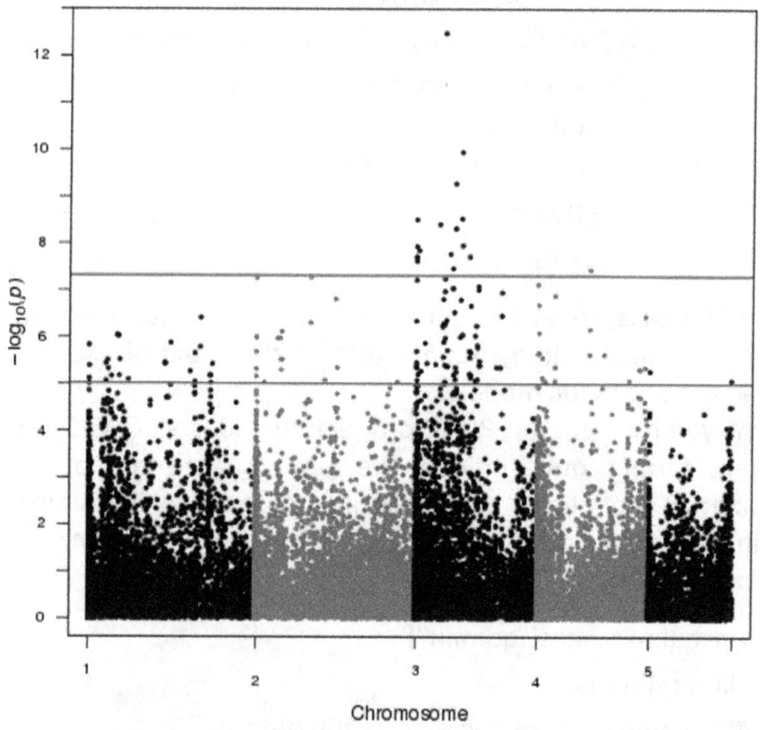

Figure 17. A Manhattan plot of uncorrected GWAS. Default suggestive line (blue) = $-\log_{10}$ (1e-5); default genome-wide line (red) = $-\log_{10}$ (1e-8).

default thresholds as suggestive line (blue; $-\log_{10}$ (1e-5)) and genome-wide line (red; $-\log_{10}$ (5e-8)).

Now, we are going to plot another one for the same GWAS analysis, but corrected using the GC model. In the *PLINK* program, the 'basic_test.assoc' file will be automatically overwritten as per Figure 14 once the '--adjust' command takes place. Take note that the corrected output does not consists of SNP position in 1.90 beta version. Hence, users need to include another column of SNP positions manually. We will continue from the previous R step for the uncorrected GWAS. The R commands are as follows:

```
>results_GC<-read.table("basic_test.assoc.adjusted", header=T)
>results_GC <-results_GC[match(as.matrix(results["SNP"]), as.matrix(results_GC["SNP"])),]
>results2 <- cbind(results[,c("SNP", "CHR", "BP")],results_GC ["GC"])
>colnames(results2)<- c("SNP", "CHR", "BP", "P")

#to color the chromosomes and set cutoff values
#to define suggestiveline and genomewideline as GC-adjusted threshold and Bonferroni-adjusted threshold
>manhattan(results2, suggestiveline = -log10(8.885e-4), genomewideline = -log10(1.99e-06),col = c("darkgreen","brown"))

#to quit R
>q()
```

After the GC correction, the number of phenotype-associated SNPs is reduced significantly as shown in a new Manhattan plot (Figure 18). Proper *p-value* thresholds should be employed in order to identify the genuine phenotype-associated SNP markers and more importantly genomic regions associated with the trait as a whole: 'QTLs'. In this case, we identify the major QTL region for our binary phenotype on chromosome 3 based on a GC-adjusted threshold (blue line). A more stringent threshold (red line) is based on Bonferroni adjustment, but no SNP marker reaches significance.

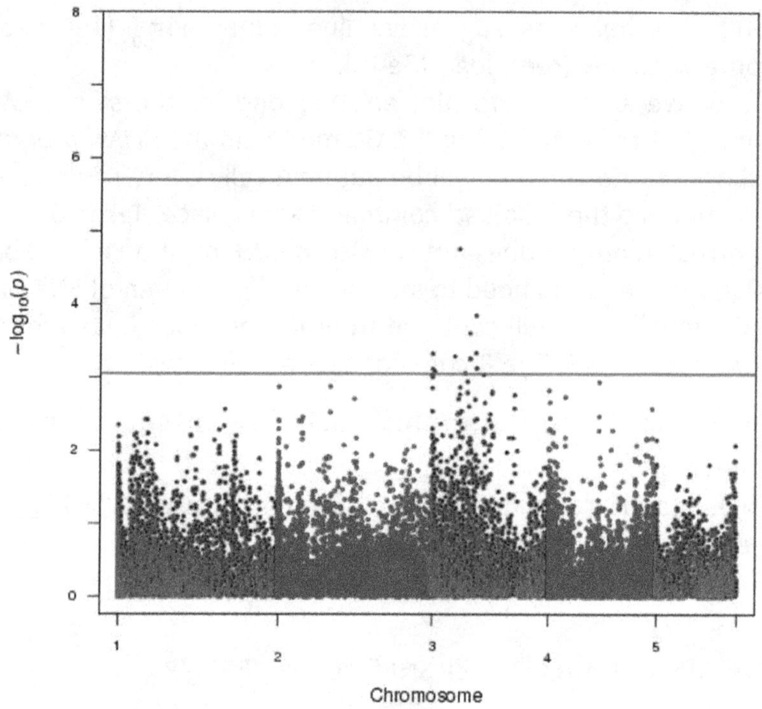

Figure 18. A colored Manhattan plot of genomic control (GC)-corrected GWAS. Bonferroni (Bonf)-adjusted threshold at the genome-wide level (red); GC-adjusted threshold (blue).

Note: The adjusted thresholds are determined in 'basic_test.assoc.adjusted' according to unadjusted *p-value* (UNADJ) at GC and Bonferroni (BONF)-adjusted *p-value* = 0.05, respectively.

The significant SNPs are those with *p-value*< 8.885e-04 can then be extracted using basic R sub-setting skills:

```
>significant.snps.selc <- results2$P < 8.885e-4
>results2.sig.snp <- results2[significant.snps.selc,]
>results2.sig.snp.sorted<-results2.sig.snp[order(result2.sig.snp$P),]
>results2.sig.snp.sorted
```

The output is sorted based on *p-values* as shown in Table 2.

Table 2. Significant phenotype-associated SNPs.

CHR	SNP	BP	P
3	snp44196	9991879	1.74E-05
3	snp79910	16196349	1.43E-04
3	snp51552	13929304	2.49E-04
3	snp79911	16198963	4.65E-04
3	snp61021	36874	4.73E-04
3	snp62103	8320415	5.14E-04
3	snp46393	13966522	5.54E-04
3	snp46394	13967604	5.54E-04
3	snp69961	16411377	7.49E-04
3	snp49079	49234	7.62E-04

CHR – Chromosome; BP – Genomic position (bp); P – Genomic control-adjusted *p-value*

By locating transcriptome sequences and GWAS profiles on the genome, trait-associated genes associated genes can be identified to understand the possible underlying biological causality, such as alternative splicing events or synonymous substitutions in the genome. In Figure 19, this is how we identify the potential genes (Gene A) for our binary trait based on the GWAS results and transcriptome locations on the genome. The most significant QTL region (yellow highlighted) is expanded in the figure and snp44196 (P_{GC} =1.74E-05) is located in the second exon of Gene A on chromosome 3. The effect of this change on Gene A, however, will need to be functionally validated. In this example, we assume that we know the genomic position of genes, which is usually the case for model species (humans, mouse, Arabidopsis etc.). For other species, how do we know whether significant SNPs occur in genic region?

Firstly, we need to have a reliable annotation of gene models and this can be done by mapping of transcriptome to the genome. This means users need to obtain transcriptome data of the species in question. Otherwise, users may try to predict gene models using

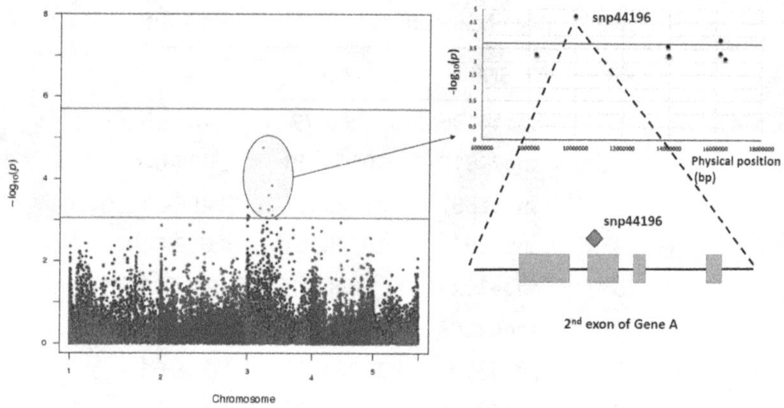

Figure 19. The potential gene (Gene A) identified in GWAS for the binary trait. The yellow-highlighted SNPs are in the association peak detected in GWAS. The snp44196 marker (a dark green diamond) is located at the second exon of Gene A.

a variety of methods such as those based on homology searching. Gene annotation is a subject of its own and it will not be further discussed here. For transcriptome mapping, we recommend the use of BLAT.[29] The program is based on a pairwise sequence alignment algorithm, much like BLAST and it was written by Jim Kent in the early 2000s. Unlike BLAST, BLAT requires very little time for whole genome scan and hence it has become an indispensable tool for genome analysis/annotation. After construction of gene models, it is now possible to find out which genes have significant phenotype-associated SNPs. On top of that, the information on gene structure potentially allows researchers to further predict the SNP effect on the function of a gene.

Summary

NGS data is massive and informative for various applications in genetic studies. By sequencing a group of individuals, researchers can now easily have access to genomic polymorphisms and then translate them to powerful DNA markers, such as SNPs and SSRs. Generating markers representative of whole genome coverage is no longer a bottleneck in linkage mapping. High-density linkage maps no doubt contribute to better QTL mapping resolution, although large population size is the most important factor for localization of QTL effects in a controlled population. More importantly, the integration of linkage maps also further allow the improvement of the genome assembly quality to construct physical reference maps, making GWAS more comprehensive. Researchers now have an opportunity to zoom in QTLs/associations and identify potential genes underlying them using transcriptome evidence. More functional studies are required, however, to confirm their causality. The pipeline has proven to be extremely useful in humans, especially in pharmacogenetics. Beneficial outcomes in animal and plant breeding programs are also foreseeable.

References

1. Mukherjee, S. *et al.* Genomes OnLine Database (GOLD) v.8: overview and updates. *Nucleic Acids Research* **49**, D723–D733 (2020).
2. Fiers, W. *et al.* Complete nucleotide sequence of bacteriophage MS2 RNA: primary and secondary structure of the replicase gene. *Nature* **260**, 500–507 (1976).
3. Oliveira, E. J., Pádua, J. G., Zucchi, M. I., Vencovsky, R. & Vieira, M. L. C. Origin, evolution and genome distribution of microsatellites. *Genetics and Molecular Biology* **29**, 294–307 (2006).
4. Rafalski, A. Applications of single nucleotide polymorphism in crop genetics. *Current Opinion in Plant Biology* **5**, 94–100 (2002).
5. Altshuler, D., Daly, M. J. & Lander, E. S. Genetic mapping in human disease. *Science* **322**, 881–888 (2008).
6. Baron, M. The search for complex disease genes: fault by linkage or fault by association? *Molecular Phychiatry* **6**, 143–149 (2001).
7. Risch, N. & Merikangas, K. The future of genetic studies of complex human diseases. *Science* **273**, 1516–1517 (1996).
8. Sturtevant, A. H. The linear arrangement of six sex-linked factors in Drosophila, as shown by their mode of association. *Journal of Experimental Zoology* **14**, 43–59 (1913).
9. Rostoks, N. *et al.* Genome-wide SNP discovery and linkage analysis in barley based on genes responsive to abiotic stress. *Molecular Genetics and Genomics* **274**, 515–527 (2005).

10. Singh, Rajinder, *et al.* "Mapping quantitative trait loci (QTLs) for fatty acid composition in an interspecific cross of oil palm." *BMC plant biology* **9**.1 (2009): 1–19.
11. Teh, C. K., Ong, A. L., Mayes, S., Massawe, F. & Appleton, D. R. Major qtls for trunk height and correlated agronomic traits provide insights into multiple trait integration in oil palm breeding. *Genes (Basel).* (2020) doi:10.3390/genes11070826.
12. Margarido, G. R. A., Souza, A. P. & Garcia, A. A. F. OneMap: software for genetic mapping in outcrossing species. *Hereditas* **144**, 78–79 (2007).
13. Lander, E. S. *et al.* MAPMAKER: an interactive computer package for constructing primary genetic linkage maps for experimental and natural populations. *Genomics* **1**, 174–181 (1987).
14. Voorrips, R. E., MapChart: Software for the graphical presentation of linkage maps and QTLs. *The Journal of Heredity* **93**(1), 77–78 (2002).
15. Tangaya, P. *et al.* A genic SNP-based high density genetic map of a Sime Darby Calix600 oil palm cross. In *Plant Genomic Congress Asia*, Kuala Lumpur (2014).
16. Ong, A. L. *et al.* Linkage-based genome assembly improvement of oil palm (*Elaeis guineensis*). *Scientific Reports* (2019) doi:10.1038/s41598-019-42989-y.
17. Wai, C. M., Moore, P. H., Paull, R. E., Ming, R. & Yu, Q. An integrated cytogenetic and physical map reveals unevenly distributed recombination spots along the papaya sex chromosomes. *Chromosome Research* **20**, 753–767 (2012).
18. Van Ooijen, J. W. JoinMap 4, Software for the calculation of genetic linkage maps in experimental populations. Kyazma B.V., Wageningen, Netherlands (2006).
19. Glazer, A. M., Killingbeck, E. E., Mitros, T., Rokhsar, D. S. & Miller, C. T. Genome assembly improvement and mapping convergently evolved skeletal traits in sticklebacks with genotyping-by-sequencing. *G3: Genes|Genomes|Genetics* **5**, 1463–1472 (2015).
20. Groenen, M. A. *et al.* A consensus linkage map of the chicken genome. *Genome Research* **10**, 137–147 (2000).
21. Matise, T. C. *et al.* A second-generation combined linkage — physical map of the human genome. *Genome Research* **17**, 1783–1786 (2007).
22. Wang, S. *et al.* Sequence-based ultra-dense genetic and physical maps reveal structural variations of allopolyploid cotton genomes. *Genome Biology* **16**, 1–18 (2015).
23. Tang, H. *et al.* ALLMAPS: Robust scaffold ordering based on multiple maps. *Genome Biology* (2015) doi:10.1186/s13059-014-0573-1.

24. Ong, A. L. *et al.* An improved oil palm genome assembly as a valuable resource for crop improvement and comparative genomics in the Arecoideae subfamily. *Plants* (2020) doi:10.3390/plants9111476.
25. Chang, C. C., Chow, C. C., Tellier, L. C. A. M., Vattikuti, S., Purcell, S. M. & Lee, J. J. Second-generation PLINK: rising to the challenge of larger and richer datasets. *GigaScience*, **4** (2015).
26. Astle, W. & Balding, D. J. Population structure and cryptic relatedness in genetic association studies. *Statististical Science* **24**, 451–471 (2009).
27. Devlin, B. & Roeder, K. Genomic Control for Association Studies. *Biometrics* **55**, 997–1004 (1999).
28. Turner, S. D. qqman: An R package for visualizing GWAS results using Q–Q and manhattan plots. *BiorXiv*, DOI: 10.1101/005165.
29. Kent, W. J. BLAT — The BLAST-Like Alignment Tool. *Genome Research* **12**, 656–664 (2002).

Chapter 11

Predicting Human Enhancers with Machine Learning

Callum MacPhillamy and Wai Yee Low

The Davies Research Centre, School of Animal and Veterinary Sciences, University of Adelaide, Roseworthy, SA 5371, Australia

Introduction

One of the most powerful methods available for bioinformaticians to use is machine learning (ML), which can be used to predict various biological features. While ML is a powerful tool, there is usually a steep learning curve and in-depth demonstration of realistic biological applications with this tool is hard to find. The aim of this chapter is to show how to predict human enhancers using a ML method and a logistic regression model to give the users a sense of how to apply them and interpret the results at the end of the practical. As this is meant for beginners with no experience in ML, the aim is not to use heavy computational resource and we downsampled the number of enhancers used in training models. For users with adequate computation resource, they can run the code on the full human genome dataset.

Enhancers are short (~50–1500 bp) non-coding regulatory sequences that increase the likelihood a gene will be transcribed.[1] They have important implications in a number of human diseases like limb deformities,[2] diabetes[3] and various cancers[4–6] and so being able to accurately identify them within the genome has important implications for how we understand the genetic controls for these diseases.

In this tutorial, we will be using a logistic regression and simple convolutional neural network (CNN) to demonstrate how one can represent DNA sequences for ML and the types of accuracy that can be achieved with increasingly non-linear models and some of the trade-offs that can occur when using more flexible models.

This tutorial is broken into three parts: (1) setting up the software environment; (2) wrangling and transforming our data; and (3) training and evaluating our models. Without further ado, let us get into it.

Setting up the Software Environment

Our first step is to set up our software environment, this ensures that any software that we already have installed does not get affected by any new software we install and vice versa. It also makes it easier to keep track of software versions. Please note, this code was written and tested using MacOS v12.2.1 and Linux Ubuntu 20.04.3 LTS and as a result many of the commands will not work on Windows.

```
$ cd # This takes us to our home directory
$ mkdir enhancer_prediction # Creates a new directory
$ cd enhancer_prediction # Enters our new directory
```

We then need to install our required software; for this, we will be using Anaconda. If you do not already have Anaconda installed, head over to https://www.anaconda.com/products/individual and follow the installation instructions.

Once you have installed Anaconda, deactivate the base environment if it is currently active. This is so we can create a new environment that we install our software into rather than into the base environment.

```
(base) $ # Appears when the base conda env is active
(base) $ conda deactivate # deactivates the base env

$ conda create -n <your_env_name> python=3.8 bedtools biopython
numpy pandas sklearn tqdm ucsc-fasize jupyter matplotlib # creates
a new env and installs the packages listed after python=3.8

$ conda activate <your_env_name> # activates your new env

$ pip install hilbertcurve # installs a new package from pip into
your env

$ pip install tensorflow==2.6
```

Note that if you encountered problems with using conda to install certain packages, you can try to, (i) install them individually one after another, (ii) use pip to install it, e.g. at the time of this writing, conda with python 3.8 will not install tensorflow, however, pip will install tensorflow 2.6 to go with python 3.8, and (iii) post the installation issues to a search engine such as DuckDuckGo to check for potential solutions.

With our software environment and directories set up we can now start downloading our data. We will be doing everything from within the directory we made at the start of this chapter so to ensure we are still where we think we are. Let us check our current directory.

```
(<your_env_name>) $ pwd
/home/enhancer_prediction
```

Now we can download the data needed to train and test our enhancer prediction model. Please note, all of this is performed with our conda environment active, I have just omitted it from the code snippets for ease of reading. The enhancers were downloaded from https://enhancer.lbl.gov/.

```
$ wget https://hgdownload.soe.ucsc.edu/goldenPath/hg19/bigZips/latest/
hg19.fa.gz

$ wget http://ftp.ensembl.org/pub/grch37/release-105/gtf/homo_sapiens/
Homo_sapiens.GRCh37.87.chr.gtf.gz

$ wget https://www.encodeproject.org/files/ENCFF001TDO/@@download/
ENCFF001TDO.bed.gz

$ gunzip *.gz

$ mv ENCFF001TDO.bed hg19-blacklist.bed

$ wget https://raw.githubusercontent.com/DaviesCentreInformatics/
Book_CHAPTER/main/hg19.VISTA.enhancers.fa?token=GHSAT0AAAAAABQIDFCB3P
WBF4DY7MTD6MAYYPPHHWA

$ mv hg19.VISTA.ehancers.fa?token=GHSAT0AAAAAABQIDFCB3PWBF4DY7MTD6MAYY
PPHHWA hg19.VISTA.enhancers.fa
```

From here on, a lot of our coding will be done using Python and Jupyter Notebooks (Figure 1), this code will have while the command line code will continue to use the style used in this chapter so far.

Figure 1. Screenshot from the browser window of the Jupyter Notebook. Select the "New" button to create a new Jupyter Notebook (.ipynb) file.

```
$ jupyter notebook # This will open the browser
# You should see the contents of the `enhancer_prediction` directory.
# Click new and create a new jupyter notebook
```

Transforming the Data

Inside the Jupyter notebook you created, click in the first cell, and type the code below. Then press shift+enter or the "Run" button while that cell is highlighted. This will run the contents of the cell and import all the necessary libraries. **Always ensure you run every code cell.**

```
# Import required libraries
import numpy as np # excellent python package for numerical computations
import pandas as pd # Python package for manipulating dataframes like .csv's. We can also
use it to handle .bed files
from Bio import SeqIO # Python package for handling biological data.
from Bio.SeqUtils import GC
import re # Python package for regular expressions
from tqdm import tqdm # Package that adds progress bars to iterators
import itertools
from sklearn.utils import shuffle # Allows you to shuffle two arrays in unison
import time
import random
import os
random.seed(12)
np.random.seed(12)
```

In a new cell below the one we just made, type the following. Feel free to ignore the comments (lines starting with #) they're just there to help explain what each part of code is doing.

```
# The VISTA enhancer set gets downloaded as a fasta file so we will have to do some
manipulating to get the genomic coordinates.
in_fa = 'hg19.VISTA.enhancers.fa' # path to the enhancer fasta file
beds = [] # empty list to append the coordinates to
print('Extracting bed regions')
```

```
for record in tqdm(SeqIO.parse(in_fa, 'fasta')):

    #print(record.name)
    rec = re.split('[|:-]', record.name) # Splits the string where ever
a '|', ':' and/or '-' occur.
    #print(rec)

    beds.append((rec[1:])) # Adds the coordinates to a list.

print('Finished extracting bed regions')

print(beds[0]) # This should print something like

# ['chr16', '86430087', '86430726']


count = 0

with open('hg19.VISTA.enhancers.chr1.bed', 'w') as bed:

    for reg in tqdm(beds):

        # Only take records that belong to chr1.

        if re.match('^chr1$', reg[0]):

            bed.write(f'{reg[0]}\t{reg[1]}\t{reg[2]}\n')

            count += 1

print(f'Finished writing chr1 enhancers. Number of enhancers written = {count}.')
```

You should have the coordinates of the 85 enhancers in BED format written to the new file. This is because we used regular expressions to take only records belonging to chromosome 1.

Next, we will look at the nucleotide composition of the enhancers. To do this we will first generate a fasta file from the bed file we just created and then we will write a python function to compute some basic statistics of the sequences in the fasta file. In the command line, type:

```
$ bedtools getfasta -fi hg19.fa -bed
hg19.VISTA.enhancers.chr1.bed -fo hg19.VISTA.enhancers.chr1.fa
# If you don't already have an index file, bedtools will create
an index of the fasta file. This only has to happen once.
```

Once this has finished go back to the Jupyter notebook and in a new cell type:

```
def fa_stats(in_fa):
    """

    Function to compute some basic descriptive stats from a
multifasta file.
    :param in_fa: Path to multifasta file
    :return: A dictionary of stats
    """
```

```
    # The main dictionary that will hold all the stats.
    stats = {}
    # The other dictionaries that store the specific stats for
length, gc and % Ns.
    lengths = {}
    gc = {}
    perc_Ns = {}
    # Iterates over the multifasta file. i = the current iteration
from the enumerate function.
    # record = the fasta record in the multifasta file.
    for i, record in tqdm(enumerate(SeqIO.parse(in_fa, 'fasta'))):
        # Get the length of the current enhancer
        lengths[f"enhancer_{i}"] = len(str(record.seq))

        # Get the GC content of the current enhancer
        gc[f"enhancer_{i}"] = GC(record.seq)

        # Get the proportion of N's in the current enhancer
        perc_Ns[f'enhancer_{i}'] = (str(record.seq).upper().
        count('N')
/ len(str(record.seq)) * 100)

    stats["Lengths"] = lengths
    stats["GC content"] = gc
    stats['Percentage of Ns'] = perc_Ns
    return stats

# Compute stats on our enhancer file.
enhancers1 = fa_stats('./enhancer_prediction/hg19.VISTA.enhancers.
chr1.fa')
# Turn the dictionary into a dataframe
enh1_df = pd.DataFrame.from_dict(enhancers1)
# Print descriptive statistics of the dataframe.
print(enh1_df.describe())
print()
# Repeat for the original (complete) enhancer set.
enhancers = fa_stats('./enhancer_prediction/hg19.VISTA.enhancers.
fa')
enh_df = pd.DataFrame.from_dict(enhancers)
print(enh_df.describe())
```

This will then print the chr1 enhancer statistics:

```
           Lengths        GC content     Percentage of Ns

count      85.000000      85.000000      85.0
mean       2109.164706    45.442165      0.0
std        1281.244669    7.084804       0.0
min        505.000000     31.382780      0.0
25%        1312.000000    39.721254      0.0
50%        1653.000000    44.746163      0.0
75%        2547.000000    51.033592      0.0
max        7787.000000    61.844569      0.0
```

As you can see the mean length of these 85 enhancer regions is quite long, with the maximum length being 7,787 bp. As we are using an input window size of 200 bp for our models, we will end up with many more than 85 enhancers to train on, as we will see next.

```
def bed_extract(in_bed, outfile, window=200, step=1, label=None,
                ignore_sex=False, append_chr=False):
    # Read in the bed file of our enhancers
    f5 = pd.read_csv(in_bed,
                     sep='\t',
                     header=None,
                     index_col=None)

    # Open a file within the `with` context to automatically close
the file once this code block finishes.
    with open(outfile,'w') as bed:
        # Iterate of the dataframe above.
        for i in tqdm(f5.itertuples()):
            chrom, start, stop = i[1:4]

            # A
            if stop - start < window:
                midpoint = start + ((stop - start) //2)
                if append_chr:
                    bed.write(f'chr{chrom}\t{midpoint - (window //
2)}\t{midpoint + (window // 2)}\t{label}\n')
                else:
                    bed.write(f'{chrom}\t{midpoint-(window//2)}\
t{midpoint+(window//2)}\t{label}\n')

            # B
            elif stop - start >= window:
                for k in range(0, ((stop - start) + 1), step):
                    if (start + k + window) <= stop:
                        if append_chr:
                            bed.write(f'chr{chrom}\t{start +
k}\t{start + k + window}\t{label}\n')
                        else:

                            bed.write(f'{chrom}\t{start+k}\
t{start+k+window}\t{label}\n')
    return f'Finished. Saved to {outfile}'
```

This function takes a bed file as input and examines each region line by line.

- # A — If it finds the region is shorter than the desired window size (default 1,000 bp), 200 bp in our case, it will find the midpoint of that region and take 200 bp centred around that midpoint. Recall from above that none of the chromosome 1 enhancers are shorter than 200 bp so it will go straight to # B.
- # B — if the region is longer than the desired window size (200 bp in our case) then it will take a sliding window through this region, taking a 200 bp window then moving 1 bp along and taking another 200 bp window until the end of the current 200 bp window is equal to the end of the current region.

We can now use this function to extract 200bp windows from our 85 chromosome 1 enhancers. The first argument is our input file, the second is our output file, window = 200 means it will extract 200 bp windows, step=1 means it will take one step along the region after each 200 bp window is extracted and label = "enhancer" will put enhancer in the 4th column of the bed file.

```
bed_extract('hg19.VISTA.enhancers.chr1.bed',
            'hg19.VISTA.enhancers.chr1.200bp.bed',
            window=200,
            step=1,
            label='enhancer')
```

Next, we need to create the non-enhancer regions. Some papers have just randomly generated DNA sequences but have been critiqued negatively by reviewers as a result. We will use the strategy that has been implemented in Min *et al.*[7] and Hong *et al.*,[8] which is to subtract all annotated regions from the genome and use the remaining unannotated regions to generate non-enhancer examples.

The first step here is to reformat the .gtf file so that it has the same format as a .bed file. There are tools one can use like BEDOPS[9] that will handle this or you can write your own function as we have done below.

```
def gtf_annotation_extract(annotation_file, output_file, append_
chr=True, chroms=list()):
    print(f'Reading in annotation file {annotation_file}')

    # Read in the gtf file.
    anno = pd.read_csv(annotation_file,
                       sep='\t',
                       comment='#',
                       header=None,
                       index_col=None)
    print(f'Writing reformatted annotations to {output_file}')
    # Open the new file to write to.
    with open(output_file, 'w') as out:

        # Iterate of the annotation dataframe.
        for i in tqdm(anno.itertuples()):
            #print(i)
            chrom, start, stop, type = i[1], i[4], i[5], i[3]
            assert type != (None or ''), 'Something\'s missing from
the type column'
            #print(chrom, start, stop, type)
            # If the gff file doesn't have chr appended to the start
of chromosomes, this part will append it; if `append_chr=True`.
            if chrom not in chroms:
                continue
            if append_chr == True:
                out.write(f'chr{chrom}\t{start}\t{stop}\t{type}\n')
            else:
```

```
            out.write(f'{chrom}\t{start}\t{stop}\t{type}\n')
    return f'Finished. Saved to {output_file}'

gtf_annotation_extract('Homo_sapiens.GRCh37.87.chr.gtf',
'Homo_sapiens.GRCh37.87.chr1.bed',
                        append_chr=True,
                        chroms=[1])
```

This function takes in a .gtf file and outputs a .bed file. We have included a parameter that appends "chr" to the start of the chromosome number as UCSC will add "chr" to the chromosome but Ensembl will not. This is important when it comes to combining the .bed files and using them with BedTools.[10] We've also included a parameter that will only output features that belong to chromosomes in the list. When we run the function, it will only write chromosome 1 features to the new file. This just makes the toy example a bit quicker when it comes to sorting the .bed files.

The new file should look like this:

```
$ head Homo_sapiens.GRCh37.87.chr1.bed
chr1    11869   14412   gene
chr1    11869   14409   transcript
chr1    11869   12227   exon
```

Now we have this done, we need to find the inferred promoter regions for each protein-coding gene. This strategy was implemented by Hong *et al.*,[8] where they define the promoter region as a 2 kb region centred around the transcription start site (TSS) of protein-coding genes.

For this next step, we will split it over two parts. First, type:

```
print(f'Reading in genome annotation file.')
genes = pd.read_csv('Homo_sapiens.GRCh37.87.chr.gtf',
                    header=None,
                    index_col=None,
                    sep='\t',
                    comment='#')

count = 0
protein_coding_idx = []
print('Finding protein-coding regions.')
for i in tqdm(genes.itertuples()):
    if re.match('gene', i[3]):
        #print(i)
        info = re.split(';',i[9])
        if 'protein_coding' in info[-2]:
            protein_coding_idx.append(i[0])
            count += 1
        #break
print(f'Number of protein coding genes in the genome: {count}')
```

This is reading in the .gtf file and scanning the third column for the word "gene", if it gets a match, it then checks the ninth column, which has a lot of information about the feature, for the text "protein_coding". If it finds a match here it then adds the row number to a list (protein_coding_idx) and adds one to the value of count. We then print the number of protein-coding genes found which is 20,327.

Secondly, type:

```
protein_df = genes.iloc[protein_coding_idx, :]

protein_df.reset_index(drop=True,
                       inplace=True)
print('Writing promoter regions to bed file.')
with open('Homo_sapiens.GRCh37.87.protein-coding.chr1.promoter_regions.
bed', 'w') as bed:
    for i in protein_df.itertuples():
        # If the chromosome isn't chromosome 1, ignore it.
        if i[1] != 1:
            continue
        if re.match('[+]', i[7]):
            bed.write(f'chr{i[1]}\t{i[4]-
1000}\t{i[4]+1000}\tpromoter_region\n')
        elif re.match('[-]', i[7]):
            assert i[7] == '-', 'Something went wrong'
            bed.write(f'chr{i[1]}\t{i[5]-
1000}\t{i[5]+1000}\tpromoter_region\n')
print('Finished writing promoter regions to file.')
```

Here, we create a new dataframe called protein_df which is made up of all the rows that matched our criteria in the previous cell, i.e. were protein-coding genes. We then open a new file to write the promoter regions to. Next, we iterate over the protein_df dataframe where we ignore features not belonging to chromosome 1. When we find a feature belonging to chromosome 1, we check what strand the gene is on. We do this to infer the direction of transcription. If the strand is "+" we assume the direction of transcription is from left to right (→) and therefore the promoter will be near the "start" coordinate in the bed file. If the strand is "—" then we assume that the direction of transcription is from right to left (←) and that the promoter will be near the "stop" coordinate in the bed file. See (https://en.wikipedia.org/wiki/BED_(file_format)) for more info about .bed file format.

We are getting close to the end of the data preparation steps I promise. Once the promoter regions are done, we need to get the blacklisted regions. These are regions that ENCODE has determined need to be excluded from analysis as they are highly repetitive, have low mapping rates or any of numerous reasons.[11] Ordinarily, you would use all the blacklisted regions as you would be using all chromosomes, however as we are only using chromosome 1, we will filter the .bed file so that we only have chromosome 1 blacklisted regions.

```
blacklist = pd.read_csv('hg19-blacklist.bed',
                        sep='\t',
                        header=None,
                        index_col=None)
with open('hg19-blacklist.chr1.bed', 'w') as bed:
    for blk in blacklist.itertuples():
        # It the region isn't on chr1, ignore it.
        if blk[1] != 'chr1':
            continue
        bed.write(f'{blk[1]}\t{blk[2]}\t{blk[3]}\t{blk[4]}\n')
```

Or if you prefer to use bash:

```
$ grep -w "^chr1" hg19-blacklist.bed | awk '{print $1"\t"$2"\
t"$3"\t"$4}' > hg19-blacklist.chr1.bed
```

Now, we can merge all these .bed files into a single .bed file of annotated regions.

```
# Merge and sort all blacklisted regions into one bed file
$ cat hg19-blacklist.chr1.bed Homo_sapiens.GRCh37.87.chr1.bed hg19.
VISTA.enhancers.chr1.200bp.bed Homo_sapiens.GRCh37.87.protein-coding.
chr1.promoter_regions.bed | sort -k1,1 -k2,2n > hg19.annotated.
regions.chr1.bed

# Find the complement of the annotated regions. These will be the
unannotated regions.
$ fasize -detailed hg19.fa | sort -k1,1 -k2,2n | bedtools complement
-L -i hg19.annotated.regions.chr1.bed -g - > hg19.unannotated.regions.
chr1.bed
```

What the second command is doing is first getting the lengths of each chromosome in hg19.fa. It is then piping this into the sort function where it will be sorted the same way as the annotated regions. It is then using bedtools complement to find the complement to the annotated regions. The addition of the –L argument in bedtools complement limits the output of bedtools complement to be only the chromosomes found in the input –i. This way we will only have the unannotated regions of chromosome 1. This helps keep our files small for this worked example. Finally, we call this new file hg19.unannotated.regions.chr1.bed.

Now we have a list of regions that are not annotated, we can generate 200 bp non-enhancer examples for training our models. We will use a variation of the bed_extract function we wrote earlier. The main variation here is that we will ignore any unannotated region smaller than our desired input window size (200 bp). This ensures we have no overlap between hg19.VISTA.enhancers.chr1.200bp.bed (positive) and hg19.VISTA.non_enhancers.chr1.200bp.bed (negative) examples as we will demonstrate below.

```
def neg_bed_extract(in_bed, outfile, window=1000, step=1,
label=None):
    neg = pd.read_csv(in_bed,
                     sep='\t',
                     header=None,
                     index_col=None)
    with open(outfile,'w') as bed:
        for i in tqdm(neg.itertuples()):
            chrom, start, stop = i[1:4]

            # If the region is shorter than 200bp, skip.
            if stop - start < window:
                continue

            elif stop - start >= window:
                #print(start, stop)
                #print(stop-start)
                for k in range(0, ((stop - start) + 1), step):
                    if (start + k + window) <= stop:

bed.write(f'{chrom}\t{start+k}\t{start+k+window}\t{label}\n')
    return f'Finished. Saved to {outfile}'

neg_bed_extract('hg19.unannotated.regions.chr1.bed',
                'hg19.VISTA.non_enhancers.chr1.200bp.bed',
                window=200,
                label='non_enhancer',
                step=100)
```

Another difference you may have noticed in how we use this function is that instead of step = 1, we use step = 100. You are welcome to play around with this but as the unannotated regions are more numerous and longer the resulting .bed file can become incredibly large. Even using just chromosome 1 with a step of 100 gives us around 1.1 million 200 bp non-enhancers.

Next, we will double check that there is no overlap between positive and negative regions with:

```
$ bedtools intersect -a hg19.VISTA.enhancers.chr1.200bp.bed -b
hg19.VISTA.non_enhancers.chr1.200bp.bed | wc -l
```

If everything has gone right, this should return 0.

Now with our positive and negative regions sorted, we can split them into training, validation, and testing sets.

```
print(f'Reading in 200bp positive regions')
pos_df = pd.read_csv(f'hg19.VISTA.enhancers.chr1.200bp.bed',
header=None, index_col=None, sep='\t')

# Remove any duplicated regions
print('Deduplicating dataframe')
pos_dedup = pos_df.drop_duplicates()

# Create an index of the rows.
print(f'Generating index of {pos_dedup.shape[0]} examples\n')
pos_df_idx = np.arange(pos_dedup.shape[0])

# Shuffle the index
print(f'Shuffling and splitting...')
shuf_idx = shuffle(pos_df_idx, random_state=12)

# Split the index into train (70%), validation (20%) and test (10%)
train_idx = shuf_idx[:int(len(shuf_idx)*0.7)]
val_idx = shuf_idx[int(len(shuf_idx)*0.7):int(len(shuf_idx)*0.9)]
test_idx = shuf_idx[int(len(shuf_idx)*0.9):]

# Using the split indices, create train, val and test splits
train_df = pos_dedup.iloc[train_idx, :]
val_df = pos_dedup.iloc[val_idx, :]
test_df = pos_dedup.iloc[test_idx, :]

print('Writing to files')
# Open three files to write the train, val and test splits to them
train_bed = open(f'hg19.VISTA.enhancers.chr1.200bp.train.bed', 'w')
val_bed = open(f'hg19.VISTA.enhancers.chr1.200bp.val.bed', 'w')
test_bed = open(f'hg19.VISTA.enhancers.chr1.200bp.test.bed', 'w')

print(f'Writing training, validation and testing sets to files for
positive 200bp to files.')

for i in tqdm(train_df.itertuples()):
    train_bed.write(f'{i[1]}\t{i[2]}\t{i[3]}\n')
for i in tqdm(val_df.itertuples()):
    val_bed.write(f'{i[1]}\t{i[2]}\t{i[3]}\n')
for i in tqdm(test_df.itertuples()):
    test_bed.write(f'{i[1]}\t{i[2]}\t{i[3]}\n')

# Close the files.
train_bed.close()
val_bed.close()
test_bed.close()
```

See the comments in the code for what each section is doing but essentially, we are taking all the enhancer regions, shuffling them, and then splitting them into training, validation, and testing splits. We then write these splits into 3 new .bed files.

We repeat this step for the negative regions.

```
print(f'Reading in negative regions')
neg_df = pd.read_csv(f'hg19.VISTA.non_enhancers.chr1.200bp.bed',
header=None, index_col=None, sep='\t')

print("Deduplicating dataframe")
neg_dedup = neg_df.drop_duplicates()

print(f'Generating index of {neg_dedup.shape[0]} examples\n')
neg_df_idx = np.arange(neg_dedup.shape[0])
print(f'Shuffling and splitting...')

shuf_idx = shuffle(neg_df_idx, random_state=12)

train_idx = shuf_idx[:int(len(shuf_idx)*0.7)]
val_idx = shuf_idx[int(len(shuf_idx)*0.7):int(len(shuf_idx)*0.9)]
test_idx = shuf_idx[int(len(shuf_idx)*0.9):]

train_df = neg_dedup.iloc[train_idx, :]
val_df = neg_dedup.iloc[val_idx, :]
test_df = neg_dedup.iloc[test_idx, :]

print('Writing to files')
train_bed = open(f'hg19.VISTA.non_enhancers.chr1.200bp.train.bed', 'w')
val_bed = open(f'hg19.VISTA.non_enhancers.chr1.200bp.val.bed', 'w')
test_bed = open(f'hg19.VISTA.non_enhancers.chr1.200bp.test.bed', 'w')
for i in train_df.itertuples():
    if re.match('chr1', i[1]):
        train_bed.write(f'{i[1]}\t{i[2]}\t{i[3]}\n')
for i in val_df.itertuples():
    if re.match('chr1', i[1]):
        val_bed.write(f'{i[1]}\t{i[2]}\t{i[3]}\n')
for i in test_df.itertuples():
    if re.match('chr1', i[1]):
        test_bed.write(f'{i[1]}\t{i[2]}\t{i[3]}\n')
train_bed.close()
val_bed.close()
test_bed.close()
```

Finally, the penultimate step. We extract our new .bed files to extract the fasta sequences from the genome.

```
# Get positive fasta
$ for i in hg19.VISTA.enhancers.chr1.200bp.*.bed; do bedtools
getfasta -fi hg19.fa -bed ${i} -fo $(basename -s .bed $i).fa; done

# Get negative fasta
$ for i in hg19.VISTA.non_enhancers.chr1.200bp.*.bed; do bedtools
getfasta -fi hg19.fa -bed ${i} -fo $(basename -s .bed $i).fa; done
```

Then we use our fa_stats function to check the GC content of the splits to ensure they are quite similar.

```
split = ['train', 'val', 'test']
for s in split:
    fa = fa_stats(f'hg19.VISTA.enhancers.chr1.200bp.{s}.fa')
    df = pd.DataFrame.from_dict(fa)
    print(df.describe())
```

The mean and standard deviation of the GC content for each split should be around 46.8 and 10.2, respectively. Now we can filter our negative sets to follow a similar GC distribution as our positive sets. As the train, validation and test split GC contents were all reasonably similar to one another, we will just use the training split's GC content as the filtering criteria for the negative splits.

```
mean = 46.862081
std_dev = 10.208668
split = ['train', 'val', 'test']

for s in split:
    print(f'Working on {s} at 200bps.')
    fa =
open(f'hg19.VISTA.non_enhancers.chr1.GCBalanced.200bp.{s}.bed', 'w')
    for record in
tqdm(SeqIO.parse(f'hg19.VISTA.non_enhancers.chr1.200bp.{s}.fa',
'fasta')):
        if re.match('chr[A-Z|a-z]', record.name):
            continue
        if (mean - std_dev) <= GC(record.seq) <= (mean + std_dev):
            SeqIO.write(record, fa, 'fasta')
    fa.close()
```

In this step, we filter each of the negative training, validation, and testing splits so that only fasta sequences with a GC content that has a mean close to 46.9 are written to the new, GCbalanced files. Then as another sanity check, we can have a look at our GC balanced fasta files to make sure their GC distribution agree with the positive sets.

```
split = ['train', 'val', 'test']
for s in split:
    fa =
fa_stats(f'hg19.VISTA.non_enhancers.chr1.GCBalanced.200bp.{s}.bed')
    df = pd.DataFrame.from_dict(fa)
    print(df.describe())
```

The mean GC value of the negative set is about 45.

Now we will use ML methods to train models to predict enhancers and evaluate their performance.

Training and Evaluating the Models

Kmer counts with Logistic Regression

Now is the fun part, where we get to train and evaluate our models. The first one we will look at is logistic regression. The input to this model will be the proportion that each kmer contributes to the sequence. We will count 1mers (A, C, G and T) 2mers (AA, AC, ... TT) and 3mers (AAA, ..., TTT). These will be our explanatory variables and the response variable is a binary category, i.e. enhancer or not enhancer.

To get started, we first write a function to count kmers in a sequence, taken from a multi fasta file. It does this by first reading the records into a dictionary, this prevents any duplicates examples from making it through as python dictionaries do not allow duplicate keys. It then scans through each record and if it finds no ambiguous bases, adds that record to a new dictionary called clean_multi_fa. If random_choice is True it will take a random sample from the clean records. This allows us to balance our datasets as we will see shortly. Next it creates a list of all possible 1mer, 2mer and 3mer combinations as well as an empty matrix to store the values as they are counted. It then reads each fasta record and counts the kmers. Each row in the matrix, kmer_mat, corresponds to a fasta record and each column (84 columns in total) corresponds to a kmer, e.g., column 1 is A and column 84 is TTT.

```
def seq2kmer(in_fa, random_choice=False, rand_n=False):
    """
    Function to convert DNA sequences to their kmer counts
    :param in_fa:
    :param random_choice:
    :param rand_n:
    :return:
    """

    print(f'Reading {os.path.basename(in_fa)} into dictionary  and removing N\'s')
    multi_fa = SeqIO.to_dict(SeqIO.parse(in_fa, 'fasta'))
    clean_multi_fa = {}
    for k, v in tqdm(multi_fa.items()):
        if 'N' not in str(v.seq).upper():
            clean_multi_fa[f'{k}'] = v
    if random_choice is True:
        random.seed(12)
        rand_clean_idx = random.sample(list(clean_multi_fa), k=rand_n)
        clean_fa = {key: clean_multi_fa[key] for key in rand_clean_idx}
        print(f'Number of clean records taken randomly: {rand_n}')
    else:
        clean_fa = clean_multi_fa
        print(f'Number of records before N removal: {len(multi_fa)}\nNumber of records
after N removal {len(clean_fa)}')

    # Creates a list of all possible 1-, 2- and 3mers.
    nuc = [''.join(n) for n in itertools.product(['A', 'C', 'G', 'T'], repeat=1)]
    nuc += [''.join(n) for n in itertools.product(['A', 'C', 'G', 'T'],repeat=2)]
    nuc += [''.join(n) for n in itertools.product(['A', 'C', 'G', 'T'],repeat=3)]
```

```
    kmer_mat = np.zeros((len(clean_fa), len(nuc)))

    print(f'Counting kmers for each record in {os.path.basename(in_fa)}')

    for n, record in enumerate(tqdm(clean_fa.items())):

        kmer_count = {f'{nu}':0. for nu in nuc}

        seq = str(record[1].seq).upper()
        # Counts all occurrences of each kmer in each sequence and gives the
proportion of its occurrences.
        for i in range(len(seq)):
            kmer_count[f'{seq[i]}'] += (1 / len(seq))

        for i in range(len(seq) - 1):
            kmer_count[f'{seq[i:i+2]}'] += (1 / int(len(seq)-1))

        for i in range(len(seq) - 2):
            kmer_count[f'{seq[i:i+3]}'] += (1 / int(len(seq)-2))

        for k, v in kmer_count.items():
            if len(k) == 1:
                v /= len(seq)
            elif len(k) == 2:
                v /= (len(seq) - 1)
            elif len(k) == 3:
                v /= (len(seq) - 2)
        for index, value in enumerate(kmer_count.items()):
            #print(index, value)
            kmer_mat[n][index] = value[1]

    return kmer_mat
```

We can now use this function on each of the fasta files we generated earlier.

```
# Positive sets
pos_train_X = seq2kmer('hg19.VISTA.enhancers.chr1.200bp.train.fa', random_
choice=False, rand_n=False)
pos_val_X = seq2kmer('hg19.VISTA.enhancers.chr1.200bp.val.fa', random_
choice=False, rand_n=False)
pos_test_X = seq2kmer('./enhancer_prediction/hg19.VISTA.enhancers.chr1.200bp.
test.fa', random_choice=False, rand_n=False)

# Negative sets
## We will use the random choice option in the function we made to take a
random sample from the negative set that is equal to the number of positive
examples we have for each split.
neg_train_X = seq2kmer('hg19.VISTA.non_enhancers.chr1.200bp.train.fa', random_
choice=True, rand_n=pos_train_X.shape[0])
neg_val_X = seq2kmer('hg19.VISTA.non_enhancers.chr1.200bp.val.fa', random_
choice=True, rand_n=pos_val_X.shape[0])
neg_test_X = seq2kmer('hg19.VISTA.non_enhancers.chr1.200bp.test.fa', random_
choice=True, rand_n=pos_test_X.shape[0])
```

Note how we set random_choice and rand_n as False for the positive examples. This is because we want to use all positive examples as we have a more limited number of validated enhancers. For the negative examples we set random_choice as True and rand_n as the same number of positive examples for the corresponding split. For example, neg_train_X will be made up of a random sample of non-enhancers from this fasta file with the number of random samples taken equal to the number of enhancers in the training set. We can then merge our training and validation sets for building a logistic regression model. This model does not have a validation step the same way our deep learning models do, which we will see in a bit.

```
# Merge the training and validation sets as logistic regression
doesn't have a validation step like the deep learning models do.
pos_train_X = np.vstack((pos_train_X, pos_val_X))
neg_train_X = np.vstack((neg_train_X, neg_val_X))

# Create labels
pos_train_y = np.ones(pos_train_X.shape[0])
pos_test_y = np.ones(pos_test_X.shape[0])
neg_train_y = np.zeros(neg_train_X.shape[0])
neg_test_y = np.zeros(neg_test_X.shape[0])

# Merge the positive and negative sets into single matrices for
training and testing
train_X = np.vstack((pos_train_X, neg_train_X))
train_y = np.hstack((pos_train_y, neg_train_y))
test_X = np.vstack((pos_test_X, neg_test_X))
test_y = np.hstack((pos_test_y, neg_test_y))
```

We then shuffle the training and testing sets, fit the model and see how well it predicts enhancers.

```
shuf_train_X, shuf_train_y = shuffle(train_X, train_y, random_state=12)
shuf_test_X, shuf_test_y = shuffle(test_X, test_y, random_state=12)

from sklearn.linear_model import LogisticRegression
log_reg = LogisticRegression(random_state=12, max_iter=1000, verbose=0)
log_reg.fit(shuf_train_X, shuf_train_y)
print(f'The the test accuracy of the logistic regression is:\t{log_reg.
score(shuf_test_X, shuf_test_y)}')
```

You will most likely get an accuracy around 67%. We can increase our confidence in this less-than-ideal performance using Kfold cross-validation.

```
# To confirm the logistic regression isn't super effective at this
task we perform cross-validation to find the mean score.
from sklearn.model_selection import KFold, cross_val_score

cross_val = KFold(n_splits=10, random_state=12, shuffle=True)
log_reg = LogisticRegression(max_iter=1000)
cv_score = cross_val_score(log_reg, shuf_train_X, shuf_train_y,
scoring='accuracy', cv=cross_val, n_jobs=-1)
print(f'The mean accuracy of the logistic regression after 10-fold
cross validation is: {np.mean(cv_score)} ({np.std(cv_score)})')
```

```
# Remove the variables to free up memory for the next model

del train_X, train_y, test_X, test_y, shuf_train_X, shuf_train_y,
shuf_test_X, shuf_test_y, neg_train_y, neg_train_X, neg_test_X, neg_
test_y, neg_val_X, pos_train_X, pos_train_y, pos_val_X, pos_test_X,
pos_test_y
```

You will most likely get a very similar score as before. This suggests that predicting enhancers from the sequence requires a deeper model than just a logistic regression. Let's try again with a deep learning model.

One-hot encoding with a simple convolutional neural network (CNN)

Onehot encoding is arguably the simplest way to represent DNA. It involves representing the DNA sequence as an $L \times 4$ matrix where L is the length of the DNA sequence and 4 is the number of nucleotides. In Figure 2, we have the 4 nucleotides along the top columns and the sequence down the rows in the first column. See how at position one, there is an A and so we place a 1 in the A column and 0's in the other columns. We then move down a row, place a 1 in the C column and 0's in the others, and so on.

For a 200 bp window size, each sequence will be represented by a matrix with shape 200×4, that is, 200 rows and 4 columns. We will now write a function that converts fasta records to one-hot matrices.

```
# Convert fasta into ML input data
def seq2onehot(in_fa, random_choice=False, rand_n=None):
    print(f'Reading {os.path.basename(in_fa)} into dictionary  and
removing N\'s')
    multi_fa = SeqIO.to_dict(SeqIO.parse(in_fa, 'fasta'))
    clean_multi_fa = {}
    for k, v in tqdm(multi_fa.items()):
        if 'N' not in str(v.seq).upper():
            clean_multi_fa[f'{k}'] = v
    if random_choice is True:
        random.seed(12)
        rand_clean_idx = random.sample(list(clean_multi_fa), k=rand_n)
        clean_fa = {key: clean_multi_fa[key] for key in rand_clean_idx}
        print(f'Number of clean records taken randomly: {rand_n}')
    else:
        clean_fa = clean_multi_fa
        print(f'Number of records before N removal: {len(multi_fa)}\
nNumber of records after N removal {len(clean_fa)}')
    seq_len = len(v.seq)
```

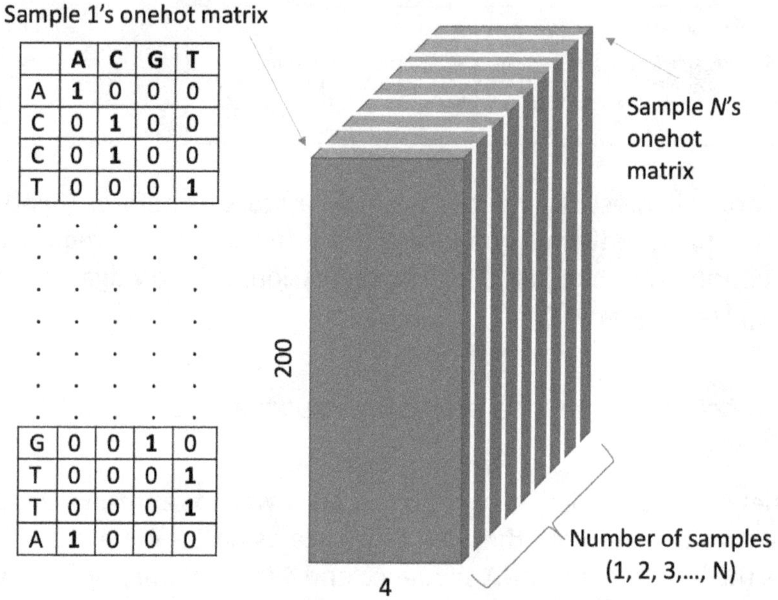

Figure 2. Graphical representation of a onehot encoded matrix of sequences. On the left, there is a table with the four nucleotides in the first row, making up each of the columns. Then in the first column, there is the sequence. As one moves along the sequence, there is a "1" in the corresponding nucleotides' column. For example, the first base is "A" so there is a "1" in the "A" column of the first row and zeroes in all others. On the right, this is a graphical representation of how each sequences' matrices would be stacked together to create a matrix of matrices or a rank 3 tensor.

```
    start_time = time.time()
    # A
    one_hot_mat = np.zeros((len(clean_fa), seq_len, 4))

    print(f'Beginning one-hot encoding of {os.path.basename(in_fa)}')
    for k, record in enumerate(tqdm(clean_fa.items())):
        for i in range(len(str(record[1].seq).upper())):
            assert str(record[1].seq).upper()[i] != 'N', 'Something
went wrong prior. You need to make sure there are no Ns in the seq'
            if str(record[1].seq).upper()[i] == 'A':
                one_hot_mat[k][i][0] = 1.
            elif str(record[1].seq).upper()[i] == 'C':
                one_hot_mat[k][i][1] = 1.
            elif str(record[1].seq).upper()[i] == 'G':
                one_hot_mat[k][i][2] = 1.
            elif str(record[1].seq).upper()[i] == 'T':
                one_hot_mat[k][i][3] = 1.
    end_time = time.time()
    total_time = end_time - start_time
    print(f'Time taken to create one-hot matrix: {total_time/60} mins')
    return one_hot_mat
```

This function works much the same way as our Kmer counting function. It differs at # A where it creates an empty matrix with shape *N, L*, 4 where *N* is the number of samples, *L* is = 200 bps and 4 is the number of nucleotides. You can think of this as a matrix of matrices or a rank 3 tensor (see Figure 2).

As the function moves along one sequence, the values in that sample's matrix are updated to make a onehot encoded sequence. As the function moves on to the next sequence, it also moves on to the next matrix in our rank 3 tensor; it repeats this until all sequences are converted.

Now that we understand how onehot encoding works, we can start converting our fasta sequences.

```
# Positive sets

pos_train_X = seq2onehot('hg19.VISTA.enhancers.chr1.200bp.train.
fa', random_choice=False, rand_n=False)
pos_val_X = seq2onehot('hg19.VISTA.enhancers.chr1.200bp.val.fa',
random_choice=False, rand_n=False)
pos_test_X = seq2onehot('hg19.VISTA.enhancers.chr1.200bp.test.fa',
random_choice=False, rand_n=False)

# Negative sets

## We will use the random choice option in the function we made to
take a random sample from the negative set that is equal to the
number of positive examples we have for each split.
neg_train_X = seq2onehot('hg19.VISTA.non_enhancers.chr1.200bp.train.
fa', random_choice=True, rand_n=pos_train_X.shape[0])
neg_val_X = seq2onehot('hg19.VISTA.non_enhancers.chr1.200bp.val.
fa', random_choice=True, rand_n=pos_val_X.shape[0])
neg_test_X = seq2onehot('hg19.VISTA.non_enhancers.chr1.200bp.test.
fa', random_choice=True, rand_n=pos_test_X.shape[0])
```

Much the same process as we did with the Kmer counts except that we do not merge the training and validation sets as Tensorflow and Keras can utilise a validation step which we will look at in a bit.

Then we need to label and stack our data together.

```
# Create the positive labels
pos_train_y = np.ones(pos_train_X.shape[0])
pos_val_y = np.ones(pos_val_X.shape[0])
pos_test_y = np.ones(pos_test_X.shape[0])

# Create the negative labels
neg_train_y = np.zeros(neg_train_X.shape[0])
neg_val_y = np.zeros(neg_val_X.shape[0])
neg_test_y = np.zeros(neg_test_X.shape[0])

# Merge the positive and negative sets into single matrices for
training and testing
train_X = np.vstack((pos_train_X, neg_train_X))
train_y = np.hstack((pos_train_y, neg_train_y))
```

```
val_X = np.vstack((pos_val_X, neg_val_X))
val_y = np.hstack((pos_val_y, neg_val_y))
test_X = np.vstack((pos_test_X, neg_test_X))
test_y = np.hstack((pos_test_y, neg_test_y))

# Shuffle the datasets.
shuf_train_X, shuf_train_y = shuffle(train_X, train_y, random_
state=12)
shuf_val_X, shuf_val_y = shuffle(val_X, val_y, random_state=12)
shuf_test_X, shuf_test_y = shuffle(test_X, test_y, random_state=12)
```

Next, we need to define our model. We will be using Tensorflow and Keras for this, however there are other deep learning libraries out there, such as PyTorch. In the code below, we create our first deep learning model. For more information on the functional API see (https://keras.io/guides/functional_api/). One of the first steps when creating a model is to set what the input size will be, deep learning models cannot take variable length input and expect everything to have the same shape, so we first set what the input shape will be which is 200,4,1. You may have noticed this is different from our previous matrix of matrices where the shape was *N*, *L*, 4 or *N*, 200, 4, recall that $L = 200$ bp. This is because in order to use a 2D convolutional layer such as those often employed by image classification models, our input needs to have 3 dimensions per sample, i.e. height, width and channel. As we have height = 200 bps and width = 4 nucleotides, we need to add the channel dimension. This does not change what is in each samples' matrix. Next, we define the conv_bn_pool_drop function. This incredibly snazzy name describes what the function is doing, it will first perform convolutions over the input data, extracting a feature map as it goes. It then normalises the output with batch normalisation; this ensures the output of the batch normalisation layer has a mean of 0 and a standard deviation of 1. This is important when training very deep networks as it speeds up training and increases model stability. These outputs are then fed into an activation layer which uses the rectified linear unit function. This function propagates positive values as they are and negative values as 0, i.e. $ReLU(x) = \max(0,x)$. While the convolution layer already has the effect of reducing the size of its output depending on the size of its stride, the max pooling

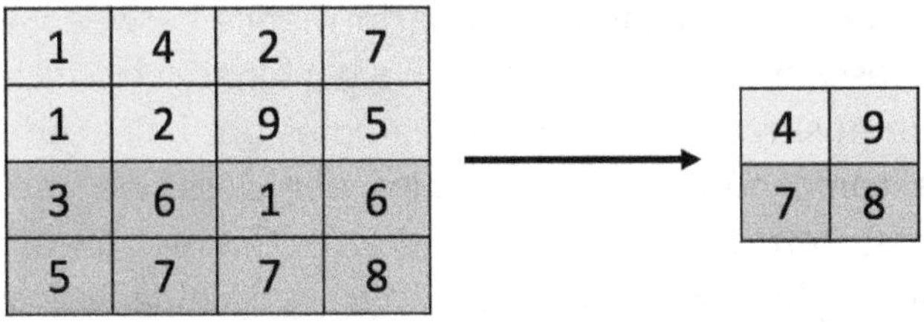

Figure 3. Graphical representation of what occurs during a max pooling operation. Here, there is a 4 × 4 matrix that we want to down sample using the max pooling operation with a window size of 2 × 2 and a stride of two. To do this, we look at the first 2 × 2 grid and see that the value "4" is the highest, we record this and move two steps to the right. We see that "9" is the highest so we record that and then move back to the left and down two. We repeat this for the whole matrix. You can see that in the resulting matrix, we have the maximum values of each of the 2 × 2 windows we examined during the max pooling operation.

layer further reduces the output size by only outputting the largest number within its pooling window, e.g. a 2 × 2 max pooling operation would look like Figure 3.

This max pooling layer can be loosely thought of as a regularisation technique in that it works to reduce the number of parameters in the model and hence the complexity. In contrast, the dropout layer is specifically employed as a regularisation technique. This randomly drops out neurons during training to introduce noise and prevents overfitting during training. Then when testing, the dropout layer is "turned off" so that the model has all neurons available for inference. We propagate our training examples through 4 of these blocks to extract features before flattening the outputs into a vector and feeding this into an artificial neural network with the final prediction being made by a sigmoid function as we just want to know if it is an enhancer or not an enhancer. We then have to construct the model where we set what the input layer is and what the output layer is; we also give it a name. Lastly, we compile the model where we set the optimizer to Adam, an optimization algorithm popular in deep learning, and the learning rate to 0.0003. Feel free to play around with these parameters, e.g. you could have a look at the difference in model performance between stochastic gradient descent and Adam. As we have a binary classification task where the positive and negative examples are balanced, i.e. they have the same number of examples per class, we can use the binary cross entropy loss function. We set the metrics to accuracy so we can see how well it performs during training and validation. Here, accuracy is $\frac{\textit{Number of correct predictions}}{\textit{Total number of predictions}}$.

```
import tensorflow as tf
from tensorflow.keras import layers

# Using the functional API, we define the model
def simpleCNN():
    # Set the input shape
    input_seq = layers.Input(shape=(200, 4, 1))
    x = layers.Reshape((200, 1, 4))(input_seq)

    # Create the feature extraction block
    def conv_bn_pool_drop(input, filters, do_rate):
        # comment in what each layer is doing.
        x = layers.Conv2D(filters=filters,
                          kernel_size=(8,1),
                          padding='same',
                          strides=(1,1),
                          use_bias=False)(input)

        x = layers.BatchNormalization()(x)
        x = layers.Activation('relu')(x)
        x = layers.MaxPooling2D(pool_size=(2,1,))(x)
        x = layers.Dropout(do_rate)(x)
        return x
    # The feature extractor
    x = conv_bn_pool_drop(x, 64, 0.1)
    x = conv_bn_pool_drop(x, 32, 0.1)
    x = conv_bn_pool_drop(x, 64, 0.)
    x = conv_bn_pool_drop(x, 64, 0.)
    # The classifier

    x = layers.Flatten()(x)
    x = layers.Dense(256, activation='relu')(x)
    x = layers.Dropout(0.5)(x)
    x = layers.Dense(128, activation='relu')(x)
    pred = layers.Dense(1, activation='sigmoid')(x)
    # Define and compile the model
    model = tf.keras.Model(inputs=input_seq,
                           outputs=pred,
                           name='SimpleCNN')
    model.compile(optimizer = tf.keras.optimizers.
Adam(learning_rate=3e-4),
                  loss = 'binary_crossentropy',
                  metrics=['accuracy'])
    return model
```

Now we can train the model. We first call the model function, then store its training cycle in the history variable. We keep batch size low as we are running this on CPU. When using a GPU, you can set batch size much larger and you will also notice a significant speed increase for training. We set the number of epochs to 10. This means we will iterate over the whole training set 10 times. We set the validation data to be our validation set and set shuffle to True, even though we have already shuffled our data.

Once this finishes training, we then plot the training and validation learning curves of the model.

```
model = simpleCNN()

history = model.fit(x = shuf_train_X,
                    y = shuf_train_y,
                    batch_size = 32,
                    epochs = 10,
                    validation_data=(shuf_val_X, shuf_val_y),
                    shuffle=True)

import matplotlib.pyplot as plt
plt.plot(history.history['accuracy'])
plt.plot(history.history['val_accuracy'])
plt.title('model accuracy')
plt.ylabel('accuracy')
plt.xlabel('epoch')
plt.legend(['train', 'validation'], loc='upper left')
plt.show()
```

The main thing to note here is that the validation accuracy and loss (orange) jumps around a bit (Figure 4(a)). There are several possible causes for this type of plot however, in our case it is likely a result of our relatively small dataset. Deep learning models thrive on large datasets and for this particular problem, it would seem that generating enhancer examples from only 85 enhancer regions of chromosome 1 does not expose the model to enough information for it to learn features associated with an enhancer. However, if you have the compute resources, you can try to repeat the work here with enhancers from all chromosomes for a more accurate prediction (Figure 4(b)).

Compare this to a model that is trained for more epochs but also on enhancers from all chromosomes. Note how much more tightly the training and validation loss follow one another. It is also worth noting that you will usually always see the training accuracy higher than the validation accuracy. However, dropout layers can lead to higher validation accuracies compared to training. As an experiment, try changing the number of epochs for our model to 25 and see what the accuracy and loss does (Figure 5).

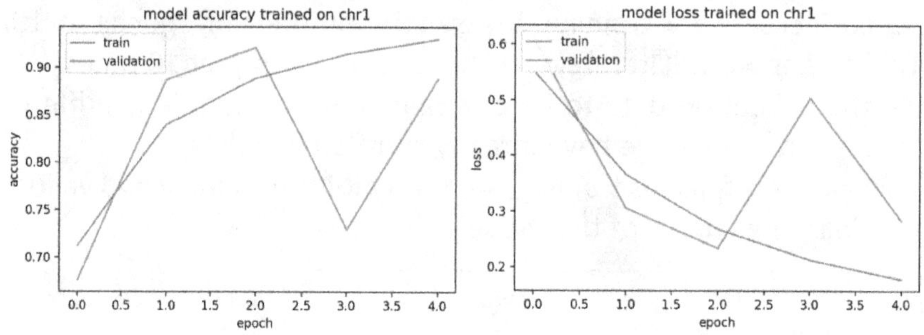

(a)

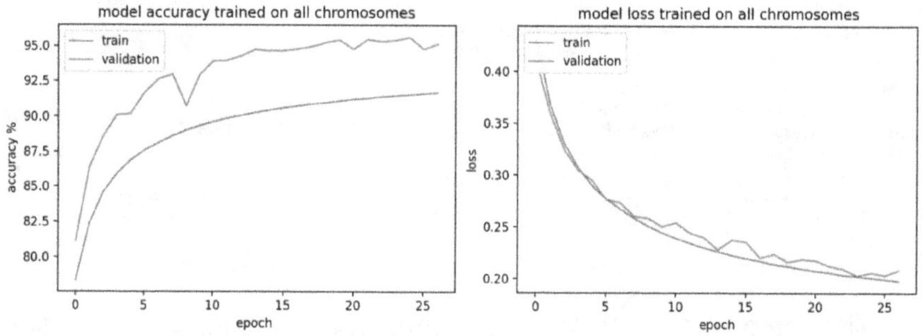

(b)

Figure 4. Comparison of accuracy and loss values when training on chromosome 1 vs all chromosomes.

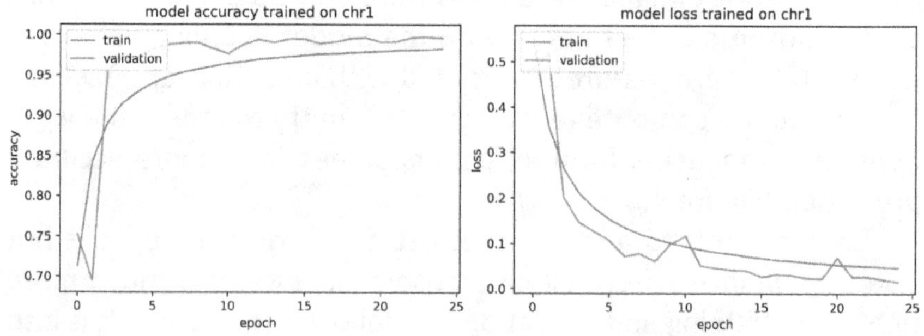

Figure 5. CNN model trained for 25 epochs on chromosome 1 data. Note the validation accuracy and loss jumping around the training score suggesting some issues with the model.

We can then evaluate the model on the previously unseen test data with

```
history.model.evaluate(shuf_test_X, shuf_test_y, verbose=1)
```

The accuracy that is printed here should be around the value of the validation accuracy for the last epoch the model performed.

Finally, we can determine how robust this performance is with Kfold cross-validation.

```
from sklearn.model_selection import KFold
kfold = KFold(n_splits=10, shuffle=True, random_state=12)
cv_scores = []
fold_no = 1
for train, val in kfold.split(shuf_train_X, shuf_train_y):
    print(f'Training on Kfold: {fold_no} of 10.')
    model = simpleCNN()
    model.fit(shuf_train_X[train], shuf_train_y[train], epochs=5)
    scores = model.evaluate(shuf_train_X[val], shuf_train_y[val])
    cv_scores.append(scores[1])
    fold_no += 1
print(f'Accuracy is: {np.mean(cv_scores)} +/- {np.std(cv_scores)}')
```

As you can see from the training and evaluation plots for the deep learning model trained on all enhancers, there is some real value to be gained from using more flexible models for enhancer prediction. A notable example of this is work by Min *et al.*[7] who trained a CNN to predict enhancers in humans using the VISTA enhancer set. With this model they were also able to accurately predict tissue specific enhancers. This highlights the potential benefits of ML in bioinformatics as it means we may now be able to start identifying more regulatory elements, like enhancers, throughout the genome without having to perform as many costly wetlab experiments to identify and then validate those regions. We can now identify candidate regions *in silico*, reducing the search space for possible enhancers.

References

1. Banerji, J., Rusconi, S. & Schaffner, W. Expression of a β-globin gene is enhanced by remote SV40 DNA sequences. *Cell* **27**, 299–308 (1981).

2. Lupiáñez, D. G. *et al.* Disruptions of topological chromatin domains cause pathogenic rewiring of gene-enhancer interactions. *Cell* **161**, 1012–1025 (2015).
3. Onengut-Gumuscu, S. *et al.* Fine mapping of type 1 diabetes susceptibility loci and evidence for colocalization of causal variants with lymphoid gene enhancers. *Nature Genetics* **47**, 381–386 (2015).
4. Helmsauer, K. *et al.* Enhancer hijacking determines extrachromosomal circular MYCN amplicon architecture in neuroblastoma. *Nature Communications* **11** (2020).
5. Northcott, P. A. *et al.* Enhancer hijacking activates GFI1 family oncogenes in medulloblastoma. *Nature* **511**, 428–434 (2014).
6. Weischenfeldt, J. *et al.* Pan-cancer analysis of somatic copy-number alterations implicates IRS4 and IGF2 in enhancer hijacking. *Nature Genetics* **49**, 65–74 (2017).
7. Min, X. *et al.* Predicting enhancers with deep convolutional neural networks. *BMC Bioinformatics* **18**, 478 (2017).
8. Hong, J., Gao, R. & Yang, Y. CrepHAN: Cross-species prediction of enhancers by using hierarchical attention networks. *Bioinformatics* **37**, 3436–3443 (2021).
9. Neph, S. *et al.* BEDOPS: High-performance genomic feature operations. *Bioinformatics* **28**, 1919–1920 (2012).
10. Quinlan, A. R. & Hall, I. M. BEDTools: A flexible suite of utilities for comparing genomic features. *Bioinformatics* **26**, 841–842 (2010).
11. Amemiya, H. M., Kundaje, A. & Boyle, A. P. The ENCODE blacklist: Identification of problematic regions of the genome. *Scientific Reports* **9**, 9354 (2019).

Index